Methods in Molecular Biology

Series Editor
John M. Walker
School of Life and Medical Sciences
University of Hertfordshire
Hatfield, Hertfordshire, UK

For further volumes:
http://www.springer.com/series/7651

For over 35 years, biological scientists have come to rely on the research protocols and methodologies in the critically acclaimed *Methods in Molecular Biology* series. The series was the first to introduce the step-by-step protocols approach that has become the standard in all biomedical protocol publishing. Each protocol is provided in readily-reproducible step-by-step fashion, opening with an introductory overview, a list of the materials and reagents needed to complete the experiment, and followed by a detailed procedure that is supported with a helpful notes section offering tips and tricks of the trade as well as troubleshooting advice. These hallmark features were introduced by series editor Dr. John Walker and constitute the key ingredient in each and every volume of the *Methods in Molecular Biology* series. Tested and trusted, comprehensive and reliable, all protocols from the series are indexed in PubMed.

Wound Regeneration

Methods and Protocols

Edited by

Hiranmoy Das

Department of Pharmaceutical Sciences, Texas Tech University Health Sciences Center, Amarillo, TX, USA

Humana Press

Editor
Hiranmoy Das
Department of Pharmaceutical Sciences
Texas Tech University Health Sciences Center
Amarillo, TX, USA

ISSN 1064-3745 ISSN 1940-6029 (electronic)
Methods in Molecular Biology
ISBN 978-1-0716-0844-9 ISBN 978-1-0716-0845-6 (eBook)
https://doi.org/10.1007/978-1-0716-0845-6

This Humana imprint is published by the registered company Springer Science+Business Media, LLC, part of Springer Nature.
The registered company address is: 1 New York Plaza, New York, NY 10004, U.S.A.

Preface

A wound is defined as a disruption in the continuity of any of the bodily tissues due to external action, typified by a cut, a bruise, a hematoma, microbial infections, or internal pathological conditions. Wounds that are not being treated or do not heal through the normal stages of healing turn out to be chronic. Comorbid pathological conditions also play a critical role in mediating chronic wounds.

The mechanisms of wound healing are as varied as the tissues in which the process is carried out. Tissue repair following injury is mediated by a number of cellular and molecular factors; failure of one or more of these components can lead to impeded healing. Because of the complex nature of wound healing, a variety of models are needed to study this process in a laboratory setting. Moreover, many novel therapeutic strategies have emerged within the field of wound regeneration, which must be evaluated extensively in preclinical models.

This book contains chapters discussing a diverse range of topics related to wound healing. The generation of wounds is used by one group as a tool to study cellular interactions and growth. Other chapters discuss the mechanisms by which different tissues regenerate. Several kinds of ischemic wounds are discussed, as well as several potential mechanisms by which these challenging wounds may potentially be treated. Several options for repairing corneal wounds are also proposed. A number of emerging technologies are proposed to help promote wound healing, including miRNA, nanomaterials, biomaterials, and stem cell therapies. Several chapters provide detailed protocols for generating wounds in animal models. As a whole, this book provides methods/protocols/information on a broad range of topics within the field of wound regeneration.

The discussed methods/protocols/information regarding wound model development and regeneration studies will be very useful in both the academic and industrial fields of research. A large range of wounds was considered for inclusion in this book. A variety of methods for in vitro, in vivo, and ex vivo studies are covered in this method book. A variety of research laboratories, students, and researchers, both from academic and industrial settings within the wound regeneration field, will find something of interest in this book.

Amarillo, TX, USA *Hiranmoy Das*

Contents

Contributors

VINAY KUMAR AAKALU • *Department of Ophthalmology and Visual Sciences, University of Illinois at Chicago, Chicago, IL, USA*

RACHEAL G. AKWII • *Department of Pharmaceutical Sciences, School of Pharmacy, Texas Tech University Health Sciences Center, Amarillo, TX, USA; Department of Pharmaceutical Sciences, Jerry H. Hodge School of Pharmacy, Texas Tech University Health Sciences Center, Amarillo, TX, USA*

SARAH ANDERSON • *Department of Pharmaceutical Sciences, Jerry H. Hodge School of Pharmacy, Texas Tech University Health Sciences Center, Amarillo, TX, USA*

JUSTINE AUSSUDRE • *CNRS UMR5287, INCIA, University of Bordeaux, Bordeaux, France*

JÉRÔME BADAUT • *CNRS UMR5287, INCIA, University of Bordeaux, Bordeaux, France; Department of Basic Sciences, Loma Linda University School of Medicine, Loma Linda, CA, USA*

DEREK BARTHELS • *Department of Pharmaceutical Sciences, Jerry H. Hodge School of Pharmacy, Texas Tech University Health Sciences Center, Amarillo, TX, USA*

BETUL N. BAYRAKTUTAR • *Center for Translational Ocular Immunology, Tufts Medical Center, Tufts University School of Medicine, Boston, MA, USA; Department of Ophthalmology, Tufts Medical Center, Tufts University School of Medicine, Boston, MA, USA; Cornea Service, Tufts New England Eye Center, Boston, MA, USA*

BRITANI N. BLACKSTONE • *Department of Materials Science and Engineering, The Ohio State University, Columbus, OH, USA*

MATILDA F. CHAN • *Department of Ophthalmology, School of Medicine, University of California, San Francisco, CA, USA; Francis I. Proctor Foundation, University of California, San Francisco, CA, USA*

TIFENN CLÉMENT • *CNRS UMR5287, INCIA, University of Bordeaux, Bordeaux, France*

HIRANMOY DAS • *Department of Pharmaceutical Sciences, Jerry H. Hodge School of Pharmacy, Texas Tech University Health Sciences Center, Amarillo, TX, USA*

COLLEEN L. DOÇI • *College of Health Professions, Marian University Indianapolis, Indianapolis, IN, USA*

PU DUANN • *Research and Development, Salem Veteran Affairs Medical Center, Salem, VA, USA*

MARIE-LINE FOURNIER • *CNRS UMR5287, INCIA, University of Bordeaux, Bordeaux, France*

CARL GREENE • *Department of Pharmaceutical Sciences, Jerry H. Hodge School of Pharmacy, Texas Tech University Health Sciences Center, Amarillo, TX, USA*

PEDRAM HAMRAH • *Center for Translational Ocular Immunology, Tufts Medical Center, Tufts University School of Medicine, Boston, MA, USA; Department of Ophthalmology, Tufts Medical Center, Tufts University School of Medicine, Boston, MA, USA; Program in Neuroscience, Graduate Biomedical Sciences, Tufts University, Boston, MA, USA; Cornea Service, Tufts New England Eye Center, Boston, MA, USA*

A K M NAWSHAD HOSSIAN • *Department of Basic Pharmaceutical and Toxicological Sciences, College of Pharmacy, University of Louisiana Monroe, Monroe, LA, USA*

ARSIA JAMALI • *Center for Translational Ocular Immunology, Tufts Medical Center, Tufts University School of Medicine, Boston, MA, USA; Department of Ophthalmology, Tufts Medical Center, Tufts University School of Medicine, Boston, MA, USA*

BRENDAN M. KENYON • *Center for Translational Ocular Immunology, Tufts Medical Center, Tufts University School of Medicine, Boston, MA, USA; Department of Ophthalmology, Tufts Medical Center, Tufts University School of Medicine, Boston, MA, USA; Program in Neuroscience, Graduate Biomedical Sciences, Tufts University, Boston, MA, USA*

MAHMOOD KHAN • *Department of Emergency Medicine, The Ohio State University Wexner Medical Center, Columbus, OH, USA; Dorothy M. Davis Heart & Lung Research Institute, The Ohio State University Wexner Medical Center, Columbus, OH, USA; Department of Physiology and Cell Biology, The Ohio State University Wexner Medical Center, Columbus, OH, USA*

RAGHURAM KOGANTI • *Department of Ophthalmology and Visual Sciences, University of Illinois at Chicago, Chicago, IL, USA*

NING XU LANDÉN • *Dermatology and Venereology Division, Department of Medicine Solna, Center for Molecular Medicine, Karolinska Institute, Stockholm, Sweden; Ming Wai Lau Centre for Reparative Medicine, Stockholm Node, Karolinska Institute, Stockholm, Sweden*

PEI-HUI LIN • *Davis Heart and Lung Research Institute, The Ohio State University, Columbus, OH, USA; Department of Surgery, The Ohio State University, Columbus, OH, USA*

XI LI • *Dermatology and Venereology Division, Department of Medicine Solna, Center for Molecular Medicine, Karolinska Institute, Stockholm, Sweden*

HANUMANTHA R. MADALA • *Department of Pharmaceutical Sciences, Jerry H. Hodge School of Pharmacy, Texas Tech University Health Sciences Center, Amarillo, TX, USA*

GEORGE MATTHEOLABAKIS • *Department of Basic Pharmaceutical and Toxicological Sciences, College of Pharmacy, University of Louisiana Monroe, Monroe, LA, USA*

CONSTANTINOS M. MIKELIS • *Department of Pharmaceutical Sciences, School of Pharmacy, Texas Tech University Health Sciences Center, Amarillo, TX, USA; Department of Pharmaceutical Sciences, Jerry H. Hodge School of Pharmacy, Texas Tech University Health Sciences Center, Amarillo, TX, USA*

ANDRÉ OBENAUS • *Department of Pediatrics, University of California, Irvine, Irvine, CA, USA; Department of Basic Sciences, Loma Linda University School of Medicine, Loma Linda, CA, USA*

GUSTAVO ORTIZ • *Center for Translational Ocular Immunology, Tufts Medical Center, Tufts University School of Medicine, Boston, MA, USA; Department of Ophthalmology, Tufts Medical Center, Tufts University School of Medicine, Boston, MA, USA; Cornea Service, Tufts New England Eye Center, Boston, MA, USA*

ARUNKUMAR PALANIAPPAN • *Department of Emergency Medicine, The Ohio State University Wexner Medical Center, Columbus, OH, USA; Centre for Biomaterials, Cell and Molecular Theranostics, Vellore Institute of Technology, Vellore, India*

PEIPEI PAN • *Department of Ophthalmology, School of Medicine, University of California, San Francisco, CA, USA*

NIKOLAUS PLESNILA • *Institute for Stroke and Dementia Research (ISD), University of Munich Medical Center, Munich, Germany*

HEATHER M. POWELL • *Department of Materials Science and Engineering, The Ohio State University, Columbus, OH, USA; Department of Biomedical Engineering, The Ohio State University, Columbus, OH, USA; Research Department, Shriners Hospitals for Children, Cincinnati, OH, USA*

Daniela Rolph • *Department of Pharmaceutical Sciences, Jerry H. Hodge School of Pharmacy, Texas Tech University Health Sciences Center, Amarillo, TX, USA*

Md S. Sajib • *Department of Pharmaceutical Sciences, Jerry H. Hodge School of Pharmacy, Texas Tech University Health Sciences Center, Amarillo, TX, USA*

Victor G. Sendra • *Center for Translational Ocular Immunology, Tufts Medical Center, Tufts University School of Medicine, Boston, MA, USA; Department of Ophthalmology, Tufts Medical Center, Tufts University School of Medicine, Boston, MA, USA*

Dhara Shah • *Department of Ophthalmology and Visual Sciences, University of Illinois at Chicago, Chicago, IL, USA*

Deepak Shukla • *Department of Ophthalmology and Visual Sciences, University of Illinois at Chicago, Chicago, IL, USA; Department of Microbiology and Immunology, University of Illinois at Chicago, Chicago, IL, USA*

Dilshan Sooriyaarachchi • *Department of Industrial, Manufacturing, and Systems Engineering, Texas Tech University, Lubbock, TX, USA*

Divya Sridharan • *Department of Emergency Medicine, The Ohio State University Wexner Medical Center, Columbus, OH, USA; Dorothy M. Davis Heart & Lung Research Institute, The Ohio State University Wexner Medical Center, Columbus, OH, USA*

Kalkunte S. Srivenugopal • *Department of Pharmaceutical Sciences, Jerry H. Hodge School of Pharmacy, Texas Tech University Health Sciences Center, Amarillo, TX, USA*

George Tan • *Department of Industrial, Manufacturing, and Systems Engineering, Texas Tech University, Lubbock, TX, USA*

Traci A. Wilgus • *Department of Pathology, The Ohio State University, Columbus, OH, USA*

Tejabhiram Yadavalli • *Department of Ophthalmology and Visual Sciences, University of Illinois at Chicago, Chicago, IL, USA*

Fatema T. Zahra • *Department of Pharmaceutical Sciences, Jerry H. Hodge School of Pharmacy, Texas Tech University Health Sciences Center, Amarillo, TX, USA*

John L. Zemkewicz • *College of Osteopathic Medicine, Marian University Indianapolis, Indianapolis, IN, USA*

Yingge Zhou • *Department of Industrial, Manufacturing, and Systems Engineering, Texas Tech University, Lubbock, TX, USA*

Chapter 1

Investigating Epidermal Interactions Through an In Vivo Cutaneous Wound-Healing Assay

John L. Zemkewicz, Racheal G. Akwii, Constantinos M. Mikelis, and Colleen L. Doçi

Abstract

Cutaneous wound healing is an intricate and multifaceted process. Despite these complexities, the distinct phases of wound healing provide a unique opportunity to evaluate the roles of different targets in these coordinated responses. This protocol details an in vivo wound healing assay to study the intersection of cellular, molecular, and systemic effector pathways. The role of certain proteins in the wound healing process can be efficiently explored in vivo through the generation of tissue-specific deficient mice. This approach, although optimized for use with animal models displaying epithelial deficiencies, can be used for other tissue-specific deficiencies, and utilizes simple and cost-effective methods, allowing investigators to precisely devise their experimental design. The coordination of immunological, epithelial, vascular, and microenvironmental factors in wound healing makes this technique a valuable tool for investigators across fields.

Key words Wound healing, Microenvironment, Epidermal

1 Introduction

Cutaneous wound healing is an intricate and multifaceted process. Healing requires the coordination of epithelial and vascular biology, signaling, stem cells, and immunological responses. Recent studies highlight the sophisticated interplay within the wound microenvironment, resulting in an increased demand for targeted methods capable of dissecting these mechanisms. Here, we present an in vivo wound-healing assay that studies the intersection of cellular, molecular, and systemic effector pathways. This protocol is optimized for use with animal models displaying epithelial deficiencies and investigates the function of certain genes in the process and the relationship between specific molecules and the wound microenvironment, making this assay a valuable tool for investigators across fields.

Hiranmoy Das (ed.), *Wound Regeneration: Methods and Protocols*, Methods in Molecular Biology, vol. 2193,
https://doi.org/10.1007/978-1-0716-0845-6_1,

One of the most robust elements of this assay resides in the ability to discern varied cellular and molecular roles of targets. These interactions are frequently absent or superficial in in vitro assays or profoundly complex and difficult to be conclusively defined in other in vivo studies. An advantage of this approach to in vivo wound healing is that it allows the investigator to evaluate their target of interest in the context of four distinct physiologic phases, each with their own discrete cellular and molecular events. These phases include hemostasis, inflammation, proliferation, and remodeling. While these phases can overlap and may vary in different genetic models, the distinct physiological processes housed within each phase can elucidate multifaceted functions of targets and their interactions [1].

Wound healing begins immediately after the skin is broken with coagulation and hemostasis and relies on the interaction between platelets, collagen, and other components of the extracellular matrix. In addition to driving clotting, the platelets also release cytokines and growth factors that stimulate vasodilation, vascular permeability, and leukocyte recruitment [2]. The second phase of the wound-healing process is the inflammatory phase, characterized by early and late stages. The early inflammatory phase initiates the innate immune response and recruits neutrophils, which in turn recruit additional inflammatory cells to the wound microenvironment [3]. Recruitment of macrophages signals the beginning of the late inflammatory phase, typically anywhere between one and three days of post-wounding. Macrophages contribute to clearing wound debris, chemokine secretion, cellular recruitment, and initiation of vascular remodeling [4]. The third phase of the healing process is the proliferative phase, which begins about 72 h after the skin is wounded and lasts between one and two weeks. The proliferative phase encompasses multiple physiological processes, including epithelialization, angiogenesis, signaling, and the proliferation and migration of epithelial cells to fill the wound [5]. Extensive remodeling of the extracellular matrix and neovascularization drive the formation of granular tissue and promote keratinocyte migration and proliferation to close the wound. The fourth phase of the wound-healing process is the remodeling phase, which involves the development of new epithelium and scar tissue formation [6]. This phase is typically initiated during the third week of wound healing but can progress for years. During remodeling, wounds increase the collagenous extracellular matrix, decrease in vascular flow, limit metabolic processes, and slow cellular growth [2].

While these phases house distinct cellular processes, successful closure and healing of wound rely on effective, coordinated mechanisms. Disruptions of critical cellular mediators in wound healing lead to readily observable defects not only in the rate of wound closure but also in the physiologic presentation of healing

[7]. Non-healing wounds, chronic ulcers, hypertrophic scars, keloids, and even tumor development have been observed upon disruption of the finely tuned balance that contributes to appropriate healing mechanisms [8–10]. These defects have also been observed in specific cell types. For example, targeted deletion of $G\alpha_q$ from epidermal cells results in defects in keratinocyte migration and differentiation and subsequently delays wound healing [11]. Similarly, loss of Rac1, CXCR2, and other inflammatory and migratory signaling factors have delayed wound healing through unique mechanisms [12–14]. Accelerated wound healing has also been observed in some transgenic models, including those with deletion of PAI-1 and tropomyosin [15, 16].

Protocols that utilize cutaneous wound healing as a tool for investigation vary greatly in the literature and can include punch biopsies, cutaneous flaps, and windows for grafting. However, many of these methods require extensive technical skills, equipment, or function best in the context of treatment interventions such as cell transplants, tissue engineering, and topical drug delivery [17–19]. The protocol presented here represents a simple and standardized approach that does not require specialized equipment and can produce robust statistical data for wound-healing alterations throughout the different phases of the process. As such, this particular iteration of the method is optimal for use in animal models with epithelial deficiencies.

2 Materials

Animal experiments should take place according to the Institutional Animal Care and Use Committee (IACUC)-approved protocols of the host institution, in compliance with the Guide for the Care and Use of Laboratory Animals. The mouse models used in any experiment may have significant impacts and can vary significantly, so investigators should carefully review their models and implement appropriate controls before starting the wound-healing protocol. This may require additional materials to those materials listed below.

2.1 Preparation of Animals for Wounding

1. Isoflurane or other inhalation anesthetic.
2. Animal hair clippers, preferably no more than 30 mm in width.

2.2 Wound Incision

1. Isoflurane inhalation anesthetic with the appropriate anesthesia induction chamber, typically regulated for 0.5–2% isoflurane and 1 L/min flow.
2. Permanent marker.
3. Ruler.
4. Iodopovidone or other topical antiseptic.

5. Scalpel #11 blade, one per three animals to be wounded (*see* **Note 1**).
6. Calipers.
7. Camera.

2.3 Wound Measurement

1. Isoflurane or other inhalation anesthetic.
2. Calipers.
3. Camera.

2.4 Analysis of Wound Closure

1. Graphing or statistical software package (*see* **Note 2**).

2.5 Preparation of Wounds for Histological Analysis

1. Materials for appropriate euthanasia.
2. Filter paper, approximately 50 mm by 70 mm strips, any weight.
3. Sharp scissors or razor blades.

3 Methods

3.1 Preparation of Animals for Wounding

1. Separate animals for use in this study. Wounded animals can be cohoused, but should not be kept with non-wounded mice (*see* **Note 3**).
2. For inducible animal models, add induction agent (i.e., tamoxifen) minimally 1 week prior to wounding (*see* **Note 4**).
3. Briefly anesthetize the animals using inhaled isoflurane.
4. Using small clippers, shave a clean rectangle of skin approximately 30 mm × 50 mm (*see* **Note 5**).
5. Return the animals to their cage and allow them to recover for 24 h.

3.2 Wound Incision (See Note 6)

1. Approximately 24 h after shaving, briefly anesthetize animals using inhaled isoflurane.
2. Carefully inspect the animals for any nicks, scratching, or other evidence of trauma and remove any animals with these signs from the experimental cohort (Fig. 1, *see* **Note 7**).
3. Using a ruler and marker, carefully draw two lines 15 mm apart on the shaved skin (Fig. 2a).
4. Liberally apply iodopovidone to the skin using a cotton swab.
5. Taking the index finger and thumb of both hands, the first investigator should pinch the skin just outside the drawn marks and roll gently back and forth (Fig. 2b, *see* **Note 8**). This rolling motion is repeated until only a thin band of the epidermis is held taut between the fingers and thumb (Fig. 2c).

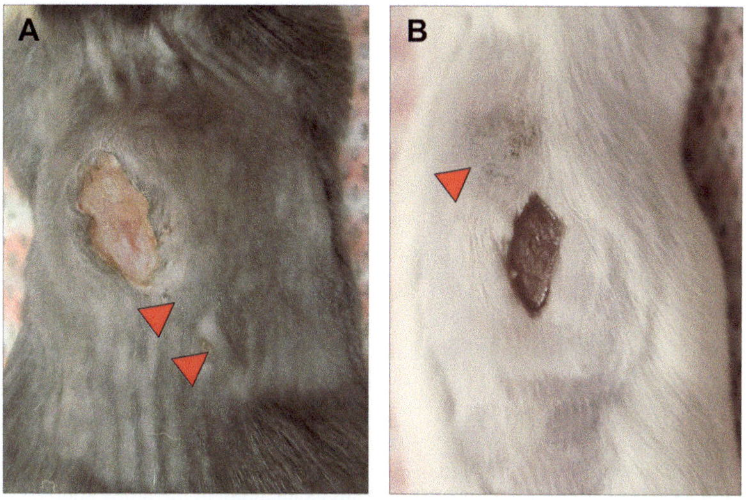

Fig. 1 Inspecting animals for injury post-shaving. Animals showing signs of trauma after shaving can include nicks (**a**, red arrowheads) and scratching or skin irritation (**b**, red arrowheads)

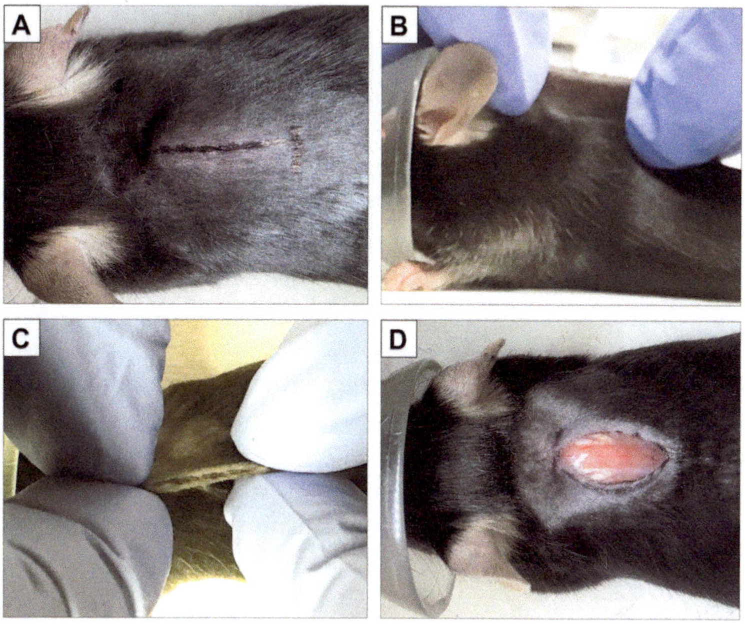

Fig. 2 Procedure for creating individual, replicable wounds. (**a**) Mice are shaved, and a 15-mm long incision line is marked. (**b**) The dorsal skin is pinched between the forefingers. (**c**) The skin is rolled repeatedly, until a full-thickness cutaneous wound without involvement of the underlying muscle layers is possible. (**d**) Appearance of a fresh wound using this approach

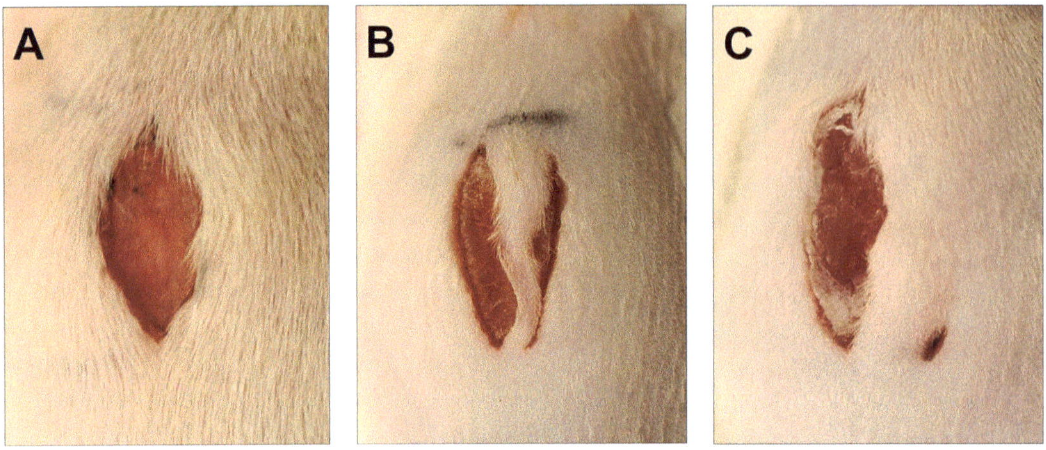

Fig. 3 Variations in wound appearance. Animals included in the study should have relatively even wounds with no additional trauma. (**a**) A clean, well-incised wound. (**b**) A double wound caused by improper rolling of the epidermis or incising with the scalpel at an angle. (**c**) A puncture wound, usually caused by slicing too deeply upon the initial incision

6. Holding a scalpel vertically, the second investigator should make a small incision that extends from one of the drawn lines to the other, exactly 15 mm (Fig. 2d, *see* **Notes 1** and **9**).
7. Release the skin and inspect the animal for any punctures, tears, or cuts outside of the principle wound (Fig. 3, *see* **Note 10**).
8. Using calipers, measure the wound length and diameter at its widest points (Fig. 4 and *see* **Note 11**).
9. Photograph each animal with a ruler for reference in each image.
10. Return the animal to its cage for recovery (*see* **Note 12**).

3.3 Wound Measurement (See Note 13)

1. Carefully anesthetize the animals (*see* **Note 14**).
2. Using calipers, measure the wound length and diameter at its widest points (Fig. 4 and *see* **Note 11**). It is recommended that the same investigator measures each time to ensure that subjective evaluations are consistent (*see* **Note 6**).
3. Photograph each animal with a ruler for reference in each image.
4. Return animals for recovery.

3.4 Analysis of Wound Closure

1. For each of the animals, calculate the total wound area as the length multiplied by the width of the wound, measured at its widest dimensions.
2. Normalize to either the measurement taken at the time of wounding or the measurement taken 12–24 h post-incision (*see* **Note 15**).

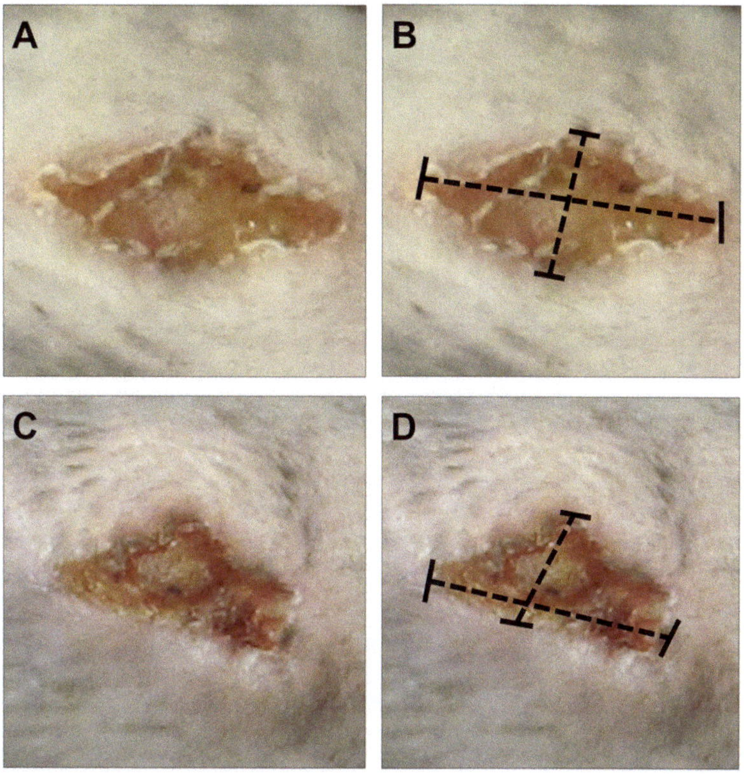

Fig. 4 Measurement of incisional wounds. Wounds can take on asymmetrical boundaries during the healing process. Establishing a consistent protocol for how wounds are measured is critical for reproducibility, and measuring at the widest dimensions is recommended. The same wound is shown in (**a**, **b**) and in (**c**, **d**). The recommended length and width measurements are indicated by the dashed line

3. Record when the wound is closed (*see* **Note 16**).
4. Investigators should consider including an analysis of wound half-life, determined from the nonlinear regression of wound closure over time (*see* **Note 2**).

3.5 Preparation of Wounds for Histological Analysis

1. Remove the mouse from the study at the desired time point for analysis (*see* **Note 14**).
2. Euthanize the mouse according to standard procedures and protocols.
3. Using sharp scissors or a razor blade, cut a square of skin with approximately 10–15 mm of normal skin adjacent to the wound (Figs. 5a, b and *see* **Note 17**).
4. Carefully place the skin with the subcutaneous layer facing down onto a piece of filter paper.

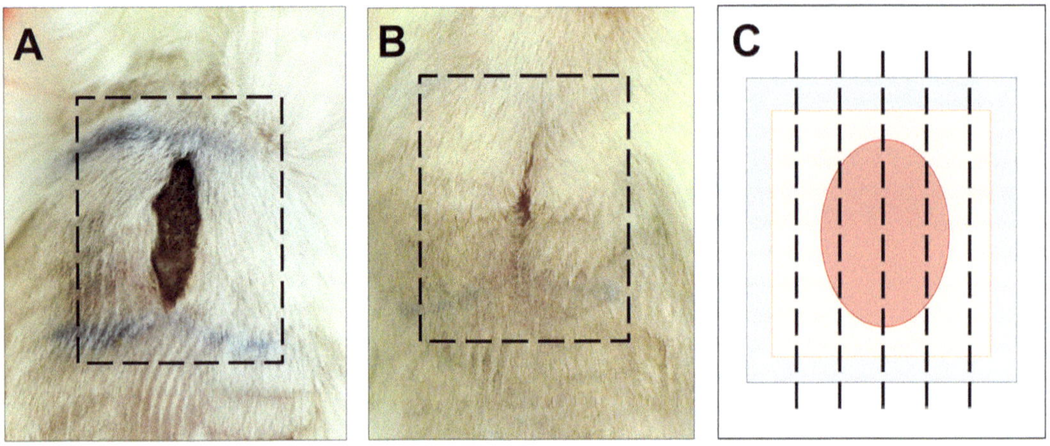

Fig. 5 Harvesting wound samples for histologic analysis. Samples from (**a**) recently wounded or (**b**) healed animals can be collected by first harvesting normal adjacent skin from wound, cutting a 30–40 mm rectangle around the wound (dashed box). (**c**) The skin should be mounted on filter paper (gray) and then cut into sagittal slices for histological processing (dashed lines)

5. Using sharp scissors or a razor blade, carefully cut the paper and skin into sagittal strips through the length of the wound (Fig. 5c and **Note 18**).
6. Proceed with appropriate histological processing for the strips.

4 Notes

1. The scalpel type depends largely on the preference of the investigator. We have found the #11 scalpel, with the elongated triangular blade, to be more accurate at making the shallow, specific cuts needed. However, the shape has less impact on the outcome than the sharpness of the blade. For that reason, we recommend using a fresh blade after every third animal.
2. For our analyses, we have used Microsoft Excel and GraphPad Prism. Other options, including SigmaPlot, R, or any other software capable of determining a nonlinear regression, are suitable.
3. Many factors can influence the selection of mice for wound-healing protocols. Some key considerations should include age, gender, and background as each can impact the mechanism and thereby the rate of wound healing. We have found that some cohorts of animals, such as the $G\alpha_{11}/G\alpha_q$ eKO previously reported [11], do not show significant differences in wound healing after approximately five months of age. Conversely, we have found in other models with genetic alterations, significant differences emerge when the animals are one year or older. Investigators have reported differences in rates of wound healing among male and female animals [20, 21]. Finally, there are

significant differences influenced by genetic background, particularly in animals whose immunogenic responses drive wound-healing mechanisms. In these cases, differences between BALB/c, C57BL6, or outbred strains such as CD1 had major consequences for healing, which should be considered. It is recommended therefore that any study using genetically modified animals should include littermate controls of the same sex.

4. The length of time for induction or treatment may vary depending on the systems impacted by the transgenic model. Full expression of the genetic variation should be confirmed in the relevant tissues before wounding.
5. Some animals have different hair density, length, and growth rates, and the average size of the shaved area should be large enough to prevent excessive hair in the wound area. For some transgenic or outbred lines, such as CD1, this may require shaving a significantly larger area than for other genotypes and backgrounds.
6. Subheading 3.2 requires two investigators. While other parts of the protocol can be done individually, it is recommended that the entirety of the technique be done in pairs to maximize efficiency and minimize the time animals spend under anesthesia. This also allows for the blinding of the study to reduce bias.
7. Some animals may be nicked during the shaving process or scratch deeply at freshly shaven skin. Animals that have these types of injuries have already initiated wound-healing mechanisms and should be excluded from the study.
8. The rolling method helps ensure the separation of the epidermal layer of skin from the muscle layers below it and is essential for ensuring that animals are wounded in an equivalent and replicable manner. When rolling, keeping the skin relatively even is important, as having it slightly twisted can lead to uneven incisions and further injuries to the mouse.
9. It is important to keep the scalpel completely vertical and make the incision as shallow as possible. This can avoid double wounding or tearing, as opposed to cutting, of the skin (Fig. 3).
10. As with shaving-induced injuries (*see* **Note** 7), additional wounds outside of the principle incision alter the wound-healing mechanisms enacted by these animals, and they should not be included in the study.
11. While good technique should result in fairly uniform incisions, this is not always the case (*see* **Notes 8–10**). Measuring at the widest and longest dimensions of the incision ensures consistent analysis of all wounds across the experimental cohort (Fig. 4).

12. Animals recover from the procedure quickly, and after a brief period, do not demonstrate signs of distress. Animals that fail to recover promptly should be evaluated by a veterinarian or member of the animal care staff.
13. Animals should be measured every 12–24 h for the first three days, after which animals can be measured every other day. Immediately after the incision, the epidermal layer will show some elasticity in closing, and the first recovery measurement will reflect a combination of skin elasticity and wound healing. The most rapid rate of wound closure will occur during the inflammatory phase, and animals should be monitored more closely during this time. Once animals have moved into the matrix remodeling and vascularization phases, the rate of wound closure decelerates, and measurements can be made less frequently.
14. The incisions often overlap regions where investigators would typically lift, immobilize, and handle the animals, so it is essential to carefully remove the animals from the cage, anesthetize the animals, and limit handling as much as possible to avoid further trauma to the incisions.
15. The normalization of the measurements is recommended but not necessary for comparison among animal cohorts. In our experience, individual differences between animals are minimized by normalizing the wound area, but the outcomes are comparable between both methods when good technique is present. Thus, it is recommended to identify each mouse included in the study and to maintain records for each individual mouse.
16. How wound closure is defined may vary between investigators. Formation of scar tissue may occur in some animal models that can make it difficult to distinguish between a closed wound and a fully healed one. We recommend using a threshold measurement as a definition of closure, often a wound area of less than 5% of the initial wound area.
17. We recommend using sharp scissors rather than razor blades to isolate and prepare the samples to avoid any tearing, but careful technique with sharp razor blades or scalpels is also appropriate.
18. Mounting the skin onto filter paper allows the investigator to evaluate multiple different cross-sections of the wound on the same slide. The thickness of the filter paper can vary and depends largely on the downstream processing of the samples. Lengthwise cuts through the tissue are essential to preserve the epithelial tongue and other histologic markers of the proliferation phase.

References

1. Martin P (1997) Wound healing--aiming for perfect skin regeneration. Science 276 (5309):75–81. https://doi.org/10.1126/science.276.5309.75
2. Velnar T, Bailey T, Smrkolj V (2009) The wound healing process: an overview of the cellular and molecular mechanisms. J Int Med Res 37(5):1528–1542. https://doi.org/10.1177/147323000903700531
3. de Oliveira S, Rosowski EE, Huttenlocher A (2016) Neutrophil migration in infection and wound repair: going forward in reverse. Nat Rev Immunol 16(6):378–391. https://doi.org/10.1038/nri.2016.49
4. Boniakowski AE, Kimball AS, Jacobs BN et al (2017) Macrophage-mediated inflammation in normal and diabetic wound healing. J Immunol 199(1):17–24. https://doi.org/10.4049/jimmunol.1700223
5. Werner S, Grose R (2003) Regulation of wound healing by growth factors and cytokines. Physiol Rev 83(3):835–870. https://doi.org/10.1152/physrev.2003.83.3.835
6. Pastar I, Stojadinovic O, Yin NC et al (2014) Epithelialization in wound healing: a comprehensive review. Adv Wound Care (New Rochelle) 3(7):445–464. https://doi.org/10.1089/wound.2013.0473
7. Wells A, Nuschke A, Yates CC (2016) Skin tissue repair: Matrix microenvironmental influences. Matrix Biol 49:25–36. https://doi.org/10.1016/j.matbio.2015.08.001
8. Schultz GS, Davidson JM, Kirsner RS et al (2011) Dynamic reciprocity in the wound microenvironment. Wound Repair Regen 19 (2):134–148. https://doi.org/10.1111/j.1524-475X.2011.00673.x
9. Linge C, Richardson J, Vigor C et al (2005) Hypertrophic scar cells fail to undergo a form of apoptosis specific to contractile collagen-the role of tissue transglutaminase. J Invest Dermatol 125(1):72–82. https://doi.org/10.1111/j.0022-202X.2005.23771.x
10. Raffetto JD (2009) Dermal pathology, cellular biology, and inflammation in chronic venous disease. Thromb Res 123(Suppl 4):S66–S71. https://doi.org/10.1016/S0049-3848(09)70147-1
11. Doci CL, Mikelis CM, Callejas-Valera JL et al (2017) Epidermal loss of Galphaq confers a migratory and differentiation defect in keratinocytes. PLoS One 12(3):e0173692. https://doi.org/10.1371/journal.pone.0173692
12. Castilho RM, Squarize CH, Leelahavanichkul K et al (2010) Rac1 is required for epithelial stem cell function during dermal and oral mucosal wound healing but not for tissue homeostasis in mice. PLoS One 5(5):e10503. https://doi.org/10.1371/journal.pone.0010503
13. Devalaraja RM, Nanney LB, Du J et al (2000) Delayed wound healing in CXCR2 knockout mice. J Invest Dermatol 115(2):234–244. https://doi.org/10.1046/j.1523-1747.2000.00034.x
14. Koo JH, Jang HY, Lee Y et al (2019) Myeloid cell-specific sirtuin 6 deficiency delays wound healing in mice by modulating inflammation and macrophage phenotypes. Exp Mol Med 51(4):48. https://doi.org/10.1038/s12276-019-0248-9
15. Chan JC, Duszczyszyn DA, Castellino FJ et al (2001) Accelerated skin wound healing in plasminogen activator inhibitor-1-deficient mice. Am J Pathol 159(5):1681–1688. https://doi.org/10.1016/S0002-9440(10)63015-5
16. Lees JG, Ching YW, Adams DH et al (2013) Tropomyosin regulates cell migration during skin wound healing. J Invest Dermatol 133 (5):1330–1339. https://doi.org/10.1038/jid.2012.489
17. Kuna VK, Padma AM, Hakansson J et al (2017) Significantly accelerated wound healing of full-thickness skin using a novel composite gel of porcine acellular dermal matrix and human peripheral blood cells. Cell Transplant 26(2):293–307. https://doi.org/10.3727/096368916X692690
18. Bello YM, Falabella AF, Eaglstein WH (2001) Tissue-engineered skin. Current status in wound healing. Am J Clin Dermatol 2 (5):305–313. https://doi.org/10.2165/00128071-200102050-00005
19. Siracusa R, Impellizzeri D, Cordaro M et al (2018) Topical application of Adelmidrol + trans-traumatic acid enhances skin wound healing in a streptozotocin-induced diabetic mouse model. Front Pharmacol 9:871. https://doi.org/10.3389/fphar.2018.00871
20. dos Santos JS, Monte-Alto-Costa A (2013) Female, but not male, mice show delayed cutaneous wound healing following aspirin administration. Clin Exp Pharmacol Physiol 40 (2):90–96. https://doi.org/10.1111/1440-1681.12043
21. Rono B, Engelholm LH, Lund LR et al (2013) Gender affects skin wound healing in plasminogen deficient mice. PLoS One 8(3):e59942. https://doi.org/10.1371/journal.pone.0059942

Chapter 2

A Murine Incisional Fetal Wound-Healing Model to Study Scarless and Fibrotic Skin Repair

Traci A. Wilgus

Abstract

The ideal response to skin injury is the complete regeneration of normal tissue without scar formation. This regenerative response is known to occur at early stages of embryonic development but is lost as the skin becomes more mature. In more developed skin, the wound-healing response is suboptimal and results in the formation of scar tissue. Scar tissue can be a significant clinical concern, causing skin dysfunction as well as psychosocial issues related to poor aesthetic outcomes. Mouse models of fetal wound healing can be useful for understanding what regulatory pathways lead to skin regeneration and scarless healing in less developed skin or scarring and fibrotic healing in more developed skin. Here, a reproducible incisional wound model in developing mice is described that our lab has used repeatedly to study scarless and fibrotic fetal wound healing.

Key words Scarless, Regeneration, Wound healing, Fetal, Skin, Development

1 Introduction

The repair of wounds in developed skin is an intricate process that involves many different cell types and regulatory mediators. This process is often divided into several stages, including hemostasis/inflammation, proliferation, and scar formation/remodeling [1–3]. The ultimate result in this situation is the production and remodeling of collagen and other extracellular matrix components by dermal fibroblasts to form scar tissue. Scar tissue acts as a quick fix to repair damage in the dermal layer of the skin, but there are several drawbacks to the production of scar tissue as opposed to the generation of normal dermal tissue. The collagen making up scar tissue is disorganized and is not aligned or assembled in the same way as normal dermal collagen, making scar tissue structurally weaker than the original tissue. In addition, scar tissue has functional deficiencies compared to normal skin, and the aesthetic issues that come with scarring can lead to a decrease in the quality of life for many patients [4].

Hiranmoy Das (ed.), *Wound Regeneration: Methods and Protocols*, Methods in Molecular Biology, vol. 2193, https://doi.org/10.1007/978-1-0716-0845-6_2,

While scar tissue is a concern with healing in mature skin, fetal skin can heal via a regenerative, scarless healing process at certain stages of development [5–7] (Fig. 1). At early gestational stages in mammals, the response to skin injury is characterized by minimal, if any, inflammation, a rapid proliferative phase with accelerated wound closure, and complete regeneration of normal skin tissue with the absence of scarring [8–10]. As fetal skin becomes more developed, it gradually loses the ability to undergo regenerative, scarless healing and begins to respond to wounds as mature skin would—with a prominent inflammatory response and subsequent scar formation (fibrotic healing). The transition from scarless to fibrotic healing takes place at around the third trimester in large animal models and in human skin [6, 11, 12].

Various animal models have been used to learn more about scarless fetal healing and about what contributes to the transition from scarless to fibrotic healing seen during development, including murine models [13–17]. Mice are a particularly useful model, as the gestation length is relatively short (18–21 days), there are abundant reagents and tools available to study mice, and transgenic/knockout mouse strains can be utilized to study the role of specific mediators or cell types in fetal wound healing in vivo. Our lab has used a murine model of incisional wound healing to study the response of fetal skin to injury for many years [10, 17–21]. In mice, studies have shown that the transition occurs at approximately embryonic day 16 (E16) [14, 17, 22, 23]. Therefore, wounds created time points before and after this age can be used to study scarless (E15) or fibrotic (E18) wound healing in fetal skin (Fig. 1).

The method outlined in this chapter has been used to generate cutaneous wounds in utero with the intention of studying the differences between scarless and fibrotic healing. Comparison of E15 (scarless) and E18 (fibrotic) wounds can be used to help define the mechanistic differences between these two distinct types of healing and to identify potential pro- or anti-fibrotic regulatory pathways. In addition, the pro-fibrotic potential of various mediators can be tested by exposing E15 wounds to the mediator in question and observing whether the wounds, which would normally heal in a scarless manner, heal with a scar. Ultimately, studying the scarless fetal wound-healing process using in vivo models could help shed light on novel strategies to enhance healing and minimize scarring.

2 Materials

2.1 Animals

1. FVB mice (or similar strain).
2. 8–10-week-old mice.

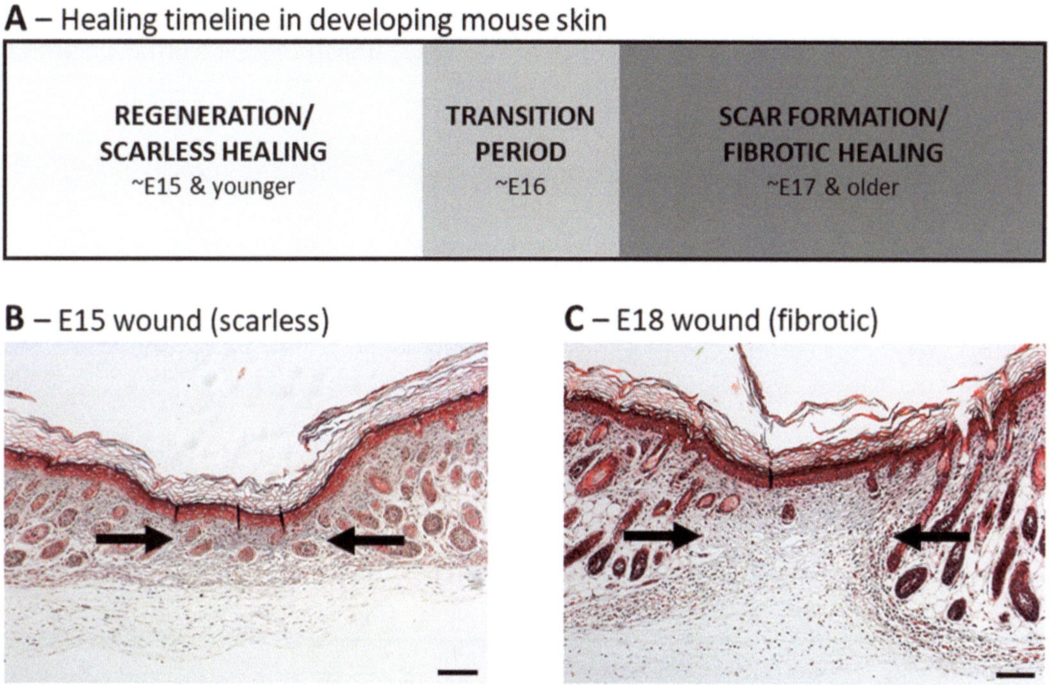

Fig. 1 Healing outcomes in murine fetal wounds. (**a**) The wound-healing response in fetal skin changes during development. In mice, the skin is able to regenerate and heals without a scar at early stages of development (until about E16). Then, there is a transition period after which skin wounds lose the ability to heal scarlessly and instead heal with a scar. (**b**, **c**) Representative Masson's trichrome-stained histological sections of 7-day wounds generated at E15 (**b**) or E18 (**c**). Note the regeneration of hair follicles within the healed wound and lack of scar tissue in E15 wounds, which is indicative of regenerative, scarless healing. Scale bar = 100 μm; arrows mark margins of the healed wound or scar

2.2 Surgical Instruments, Equipment, and Materials

1. Forceps (fine-tipped, toothed, curved, and blunt-end forceps may be needed).
2. Scissors.
3. Microsurgical scissors.
4. Needle holder.
5. Hamilton syringe (10 μl), with removable needle.
6. Gauze (nonwoven).
7. Autoclave pouches.
8. Isothermal heating pad (or other heat source).
9. 3-ml syringe with needle, sterile.
10. Anesthetic machine for small animals equipped with oxygen tank/regulator and isoflurane vaporizer (or other method of anesthesia).
11. Dissecting microscope.
12. Light source (gooseneck microscope illuminator).

13. Absorbent pads.
14. Suture (6-0 polypropylene, nonabsorbable suture with C-1 taper needle, or similar).
15. Clippers.
16. Surgical personal protective equipment (gown, mask, hair bonnet).
17. Sterile surgical gloves.
18. Cotton tip applicators.

2.3 Reagents

1. Sterile saline (0.9% sodium chloride, for injection).
2. India ink.
3. Isoflurane.
4. Chlorhexidine scrub (4% w/v).
5. 70% ethanol.
6. Xylazine.

3 Methods

Before any studies using mice are performed, the institutional animal care and use committee should be consulted. Official approval of the protocols and procedures should be obtained prior to conducting studies.

3.1 Generation of Timed Pregnant Mice

1. Prior to breeding, arrange mice so that they are housed four mice per cage for females and one mouse per cage for males. The number of cages will depend on the number of wounded fetuses desired. Mice should be a minimum of 8–10 weeks of age for optimal breeding results.
2. Once per week, place female mice from a single cage into the cage of a designated male. The following morning, check each female mouse for the presence of a vaginal plug, record the results, and return the female mice to their original cage (*see* **Note 1**). The day of plug detection is designated E0, which should be used to calculate the appropriate dates for surgery (E15 or E18 for scarless or fibrotic healing, respectively) (*see* **Note 2**). Repeat this mating procedure weekly on the same day of the week, until pregnant mice are obtained.
3. At least weekly, check females for signs of pregnancy. This is usually obvious based on weight gain and the appearance of a bulging abdomen. Palpation can also be used if needed. For surgeries being performed at E15 or older, typically visual signs of pregnancy are apparent several days prior to the surgery date.

4. At the desired embryonic age, prepare pregnant female mice for surgery as described below. Continue mating any remaining female mice that are not pregnant as described above.

3.2 Presurgical Preparation

1. Prior to surgeries, sterilize surgical instruments and other objects/reagents that will be used during surgery.
2. If an isothermal pad will be used as a heat source, preheat the pad by microwaving or submerging in a heated water bath as directed by the manufacturer.
3. Dilute India ink to 10% in sterile saline and dilute or prepare xylazine and other drugs/experimental reagents that will be used prior to, during, or after surgery. To maintain sterility, this should be performed in a biosafety hood. Load the Hamilton syringe with diluted India ink (*see* **Note 3**). Using aseptic technique, fill several 3-ml syringes with sterile saline (these will be used to keep the abdomen hydrated during surgery).
4. If inhaled isoflurane will be used for anesthesia, prepare the anesthetic machine. Fill the anesthetic chamber in the vaporizer with isoflurane. Turn on the oxygen tank to ensure that there are sufficient oxygen levels to last for the duration of the surgery.
5. Prepare the surgical space. Arrange the dissecting scope, light source, and anesthetic machine appropriately. Disinfect surfaces with 70% ethanol. Place a layer of absorbent padding on the surgical surface, followed by the preheated isothermal pad, then another layer of absorbent padding.
6. Prepare the animal for surgery. Induce and then maintain anesthesia with a suitable level of isoflurane. Appropriate depth of anesthesia should be confirmed by lack of a toe-pinch response. Using the lowest dose of isoflurane that will maintain appropriate anesthesia depth will minimize the risk of complications. Shave the abdomen of the mouse, then move the shaved mouse to the surgical surface, arranging the animal in the middle of the warm isothermal pad. After shaving the mouse, the surgeon should wear a gown, face mask, and hair bonnet. Disinfect the abdominal skin with chlorhexidine scrub followed by 70% ethanol using soaked cotton tip applicators. Use a circular motion starting at the middle of the abdomen and moving outward. Repeat the chlorhexidine and ethanol washes three times.
7. Open a packet of surgical gloves, sutures, and surgical pack, which should contain sterilized instruments and gauze. Avoid contacting the inner contents with nonsterile items from this point forward. The sterile 3-ml syringes loaded with saline should also be added. Carefully, put on sterile surgical gloves without contaminating the working surfaces of the gloves.

Arrange four pieces of sterile gauze on the abdomen of the mouse to create a surgical drape, leaving a rectangular space in the middle large enough for the laparotomy incision to be made. Use additional sterile gauze to cover any surfaces that may need to be touched or adjusted during the surgery (light source, isoflurane gauge, microscope controls, etc.). Once the animal is prepared, adjust the light source and scope if necessary.

3.3 Surgical Procedures

1. Perform a midline laparotomy. Using forceps, grasp the abdominal skin and create a vertical incision along the midline using scissors. The incision should be about 4 cm in length. In a similar manner, create an incision in the abdominal muscle wall. Toothed forceps can be used. Cutting carefully along the linea alba reduces the chance of bleeding.
2. With closed curved forceps or a needle holder, expose a portion of the uterus on one side of the animal (there will be a uterine segment on each side of the mouse) by scooping the instrument under the uterus and gently lifting. Try to minimize the amount of the uterus exteriorized. Looking through the dissecting microscope, identify an embryo that is situated such that it can be seen clearly through the scope and is in a position allowing for easy access to the dorsal skin. In our experience, it is best to avoid the two fetuses directly adjacent to the cervix on either side. From this point forward during the surgery, the exteriorized portion of the uterus should be sprayed periodically with small volumes of sterile saline to maintain hydration.
3. Taking care to avoid cutting any visible vessels in the uterus, make a small incision (3–4 mm) through the uterine wall using microsurgical scissors. Then, cut through the amniotic membranes. This should result in visible leakage of amniotic fluid, especially at earlier gestational ages.
4. Using fine-tipped forceps, gently grip the dorsal skin in the middle of the fetus and make a small (approximately 2 mm) full-thickness incision using microsurgical scissors. Care should be taken to avoid damaging the underlying tissues.
5. Using the Hamilton syringe, inject 1 μl of 10% India ink solution subcutaneously at the edges of the wound. This allows for identification of the wound site after healing has occurred. Spray the area with saline.
6. Close the incision in the uterus with a horizontal mattress suture.
7. Repeat the wounding process on several fetuses. Limiting the number of wounded fetuses to four (two on each side of the uterus) and minimizing the total time of the surgery reduce the risk of complications. Once the desired number of fetuses has

been wounded, carefully replace the uterus without introducing torsion.

8. Close the laparotomy incisions in the pregnant female mouse using multiple simple interrupted sutures, first for the abdominal wall and then for the skin (*see* **Note 4**).

3.4 Postoperative Care and Tissue Harvest

1. Immediately after surgery, inject xylazine (3 mg/kg) subcutaneously for sedation and analgesia or administer other postoperative analgesics.
2. Return the mouse to its cage, laying it on gauze. Place the isothermal pad under the cage. Monitor the mouse every 30 min for the first 3 h. Then, remove the gauze from the cage and return the cage to its housing location. Monitor the mouse at least once per day until wounds are harvested.
3. At the desired time point post-injury, euthanize the female mouse in accordance with American Veterinary Medical Association Guidelines. Depending on the postsurgical time point and the embryonic age at which the surgeries were performed, wounds need to be harvested either from unborn embryos or from recently delivered pups. In either case, decapitate the embryos or postnatal mice with scissors prior to harvesting skin samples (*see* **Note 5**).

4 Notes

1. Since the presence (or absence) of a vaginal plug does not always accurately predict pregnancy, we separate the female and male mice after overnight mating and limit mating to once per week. If natural breeding is not effectively resulting in pregnancies or if a large number of pregnant mice are needed, superovulation may be considered [13].
2. In our experience, it is difficult to create wounds on embryos younger than E15 because the skin is less developed, gelatinous, and not firm enough to manipulate with forceps or create a well-defined incision.
3. Mediators (lipids, recombinant proteins, chemicals, etc.) can be added to the India ink solution and injected into E15 or E18 wounds to evaluate pro- or anti-fibrotic effects, respectively.
4. Closing the abdominal wall and skin with multiple interrupted sutures rather than one running suture minimizes the risk of the incisions reopening if the mouse chews through the suture material.
5. For harvesting wound tissue at early time points post-wounding for early embryonic wounds (i.e., E15), the skin is quite thin and fragile. This can make dissecting the skin/

wound away from underlying tissues without damaging it difficult. If the tissue is to be fixed for subsequent histology, it may be easier to fix the entire decapitated embryo or a large section of skin with underlying tissues attached, and then separate the skin post-fixation when it is more firm.

Acknowledgments

The Wilgus lab is supported in part by NIH grant AR071115.

References

1. Martin P (1997) Wound healing--aiming for perfect skin regeneration. Science 276 (5309):75–81
2. Gurtner GC, Werner S, Barrandon Y et al (2008) Wound repair and regeneration. Nature 453(7193):314–321. https://doi.org/10.1038/nature07039
3. Eming SA, Martin P, Tomic-Canic M (2014) Wound repair and regeneration: mechanisms, signaling, and translation. Sci Transl Med 6 (265):265sr266. https://doi.org/10.1126/scitranslmed.3009337
4. Brown BC, McKenna SP, Siddhi K et al (2008) The hidden cost of skin scars: quality of life after skin scarring. J Plast Reconstr Aesthet Surg 61(9):1049–1058. https://doi.org/10.1016/j.bjps.2008.03.020
5. Armstrong JR, Ferguson MW (1995) Ontogeny of the skin and the transition from scar-free to scarring phenotype during wound healing in the pouch young of a marsupial, *Monodelphis domestica*. Dev Biol 169(1):242–260
6. Longaker MT, Whitby DJ, Adzick NS et al (1990) Studies in fetal wound healing, VI. Second and early third trimester fetal wounds demonstrate rapid collagen deposition without scar formation. J Pediatr Surg 25 (1):63–68; discussion 68–69
7. Lorenz HP, Whitby DJ, Longaker MT et al (1993) Fetal wound healing. The ontogeny of scar formation in the non-human primate. Ann Surg 217(4):391–396
8. Liechty KW, Adzick NS, Crombleholme TM (2000) Diminished interleukin 6 (IL-6) production during scarless human fetal wound repair. Cytokine 12(6):671–676
9. Liechty KW, Crombleholme TM, Cass DL et al (1998) Diminished interleukin-8 (IL-8) production in the fetal wound healing response. J Surg Res 77(1):80–84
10. Wulff BC, Parent AE, Meleski MA et al (2012) Mast cells contribute to scar formation during fetal wound healing. J Invest Dermatol 132 (2):458–465. https://doi.org/10.1038/jid.2011.324
11. Lorenz HP, Longaker MT, Perkocha LA et al (1992) Scarless wound repair: a human fetal skin model. Development 114(1):253–259
12. Rowlatt U (1979) Intrauterine wound healing in a 20 week human fetus. Virchows Arch A Pathol Anat Histol 381(3):353–361
13. Walmsley GG, Hu MS, Hong WX et al (2015) A mouse fetal skin model of scarless wound repair. J Vis Exp 95:52297. https://doi.org/10.3791/52297
14. Stelnicki EJ, Bullard KM, Harrison MR et al (1997) A new in vivo model for the study of fetal wound healing. Ann Plast Surg 39 (4):374–380
15. Naik-Mathuria B, Gay AN, Yu L et al (2008) Fetal wound healing using a genetically modified murine model: the contribution of P-selectin. J Pediatr Surg 43(4):675–682. https://doi.org/10.1016/j.jpedsurg.2007.12.007
16. Liechty KW, Kim HB, Adzick NS et al (2000) Fetal wound repair results in scar formation in interleukin-10-deficient mice in a syngeneic murine model of scarless fetal wound repair. J Pediatr Surg 35(6):866–872; discussion 872–863
17. Wilgus TA, Bergdall VK, Tober KL et al (2004) The impact of cyclooxygenase-2 mediated inflammation on scarless fetal wound healing. Am J Pathol 165(3):753–761
18. Dardenne AD, Wulff BC, Wilgus TA (2013) The alarmin HMGB-1 influences healing outcomes in fetal skin wounds. Wound Repair Regen 21(2):282–291. https://doi.org/10.1111/wrr.12028

19. Wilgus TA, Bergdall VK, Dipietro LA et al (2005) Hydrogen peroxide disrupts scarless fetal wound repair. Wound Repair Regen 13 (5):513–519
20. Wilgus TA, Oberyszyn TM, Bergdall VK et al (2005) VEGF-associated angiogenesis promotes cutaneous scar tissue deposition. FASEB J 19(4):A482
21. Wulff BC, Pappa NK, Wilgus TA (2019) Interleukin-33 encourages scar formation in murine fetal skin wounds. Wound Repair Regen 27(1):19–28. https://doi.org/10.1111/wrr.12687
22. Whitby DJ, Ferguson MW (1991) Immunohistochemical localization of growth factors in fetal wound healing. Dev Biol 147(1):207–215
23. Whitby DJ, Ferguson MW (1991) The extracellular matrix of lip wounds in fetal, neonatal and adult mice. Development 112(2):651–668

Chapter 3

Development of Cutaneous Wound in Diabetic Immunocompromised Mice and Use of Dental Pulp–Derived Stem Cell Product for Healing

Carl Greene and Hiranmoy Das

Abstract

Chronic nonhealing wounds impact nearly 15% of Medicare beneficiaries (8.2 million) in the United States costing \$28–\$32 billion annually. Despite advancement in wound management, approximately 8% of diabetic Medicare beneficiaries have a foot ulcer and 1.8% will have an amputation. The development of a regenerative approach is warranted to save these before-mentioned amputations. To this extent, herein, we describe the detailed methods in generating a type 1 diabetes mellitus (T1DM) condition in immunocompromised mice, inducing cutaneous wound, and application of dental pulp stem cell-derived secretory products for therapeutic assessment. This model helps in evaluating the efficacy of stem cell-based therapy and helps with the investigation of involved mechanisms in impaired cutaneous wound healing caused by hyperglycemic stress due to type 1 diabetes.

Key words Type 1 diabetes mellitus, Dental pulp–derived stem cell, NOD/SCID mice, Cutaneous wound

1 Introduction

Cutaneous wound healing involves the differentiation, migration, proliferation, and apoptosis of various cell types in a process recapitulating embryonic skin development [1]. Wounds failing to achieve reepithelialization in a sequential and timely manner are considered to be chronic nonhealing wounds and have an impact of almost \$32 billion annually on the U.S. healthcare system alone. The complexity involved with cutaneous wound healing and its impairment in individuals with various diseases gives rise to an urgent clinical need for new advanced wound-healing therapies in the clinical setting [2–6]. An example of pathological conditions that impair cutaneous healing would be type 1 diabetes mellitus (T1DM). Diabetic wound infections have the second-highest prevalence with surgical infections being the most common. It has been estimated that in developed countries, chronic nonhealing wounds

Hiranmoy Das (ed.), *Wound Regeneration: Methods and Protocols*, Methods in Molecular Biology, vol. 2193,
https://doi.org/10.1007/978-1-0716-0845-6_3,

have a prevalence rate of 1–2% within the general population [7, 8], which is similar to the prevalence rate for heart failure [9]. Persons suffering from nonhealing cutaneous wounds are likely to be older adults who are nonambulatory or paralyzed, making them less able to provide for self-care. Their nonhealing ulcers are thus often the result of a unique and underlying medical condition (e.g., sickle cell anemia) [10, 11] in association with renal impairment (e.g., calciphylaxis) [12], immunosuppression (e.g., steroid use) [13, 14], and autoimmune diseases (e.g., systemic lupus erythematosus). Additionally, age-related debility or paralysis can lead to pressure ulcers [15, 16]; diabetic peripheral neuropathy can result in diabetic foot ulcers (DFUs) and lastly patients with peripheral arterial and venous disease (e.g., arterial and venous ulcers).

The chronic nonhealing wound type is not so much a disease, in of itself, as it is a symptom of other underlying illnesses such as the before-mentioned type 1 diabetes mellitus (T1MD). Despite differences in etiology, at the molecular level chronic wounds share certain common features including excessive levels of proinflammatory cytokines, proteases, and ROS. Prolonged oxidative stress leads to DNA damage-related cell cycle arrest resulting in a senescent cell population [17]. There also will be the existence of persistent infection and a deficiency of stem cells that are often also dysfunctional. Briefly, $CD90^+$ mesenchymal stem cells (MSCs) are multipotent progenitor cells that can differentiate into mesenchymal tissues, such as cartilage, tendon, and bone [18]. Through their paracrine (trophic activity), they have the ability to affect the cutaneous repair process by synthesizing various essential growth factors and cytokines that have a direct effect on cell migration, proliferation, and metabolic activity of affected cells and tissues [19]. $CD90^+$ cell secretory products play an active role in the inflammatory, proliferative, and remodeling phases of wound healing [20]. In this section, we specifically focus on the step-by-step methodology in the generation of a diabetic condition in immunodeficient mice and the therapeutic application of secretory products derived from $CD90^+$ dental pulp stem cells (DPSCs) to treat cutaneous wounds on before-mentioned mice. This topical application of stem cell product helps normalizing nonhealing wound bed through improving cytokine profile in favor of healing. If successful in clinical testing, this method of application will not require any stem cell transplantation for chronic wound therapy, avoiding potential immune rejection.

2 Materials

All experiments descried here within are to be performed using aseptic techniques in sterile conditions using an approved BSL2 laboratory facility (*see* **Note 1**). Reagents are to be prepared at room temperature (*see* **Notes 2** and **3**). All procedures to be performed on rodents need prior protocol approval from the Institutional Animal Care and Use Committee (IACUC). Discard all waste materials in accordance with biohazard disposal regulations (*see* **Notes 3**, **5**, and **10**).

2.1 Generation of Type 1 Diabetes Mellitus (T1DM) in Mice

1. Nonobese diabetic/severe combined immunodeficiency (NOD/SCID): ~25 g, 8–12 weeks old.
2. Standard rodent chow diet.
3. 50 mM sodium citrate buffer, pH 4.5: prepared immediately before use.
4. Streptozotocin (STZ).
5. Rodent cages: temperature-, humidity-, and light-controlled housing.
6. "One Touch Basic" blood glucose monitoring system.
7. 1.5-mL microcentrifuge tubes.
8. Sterile biosafety cabinet level-2 (BSL-2).
9. Sterile injection needles.
10. Weight scale.

2.2 Generation of Cutaneous Wound

1. Razor.
2. Hair removing cream.
3. Betadine.
4. Biopsy tissue punch.
5. Surgical scalpel.
6. Anesthesia machine.
7. Heat blanket.

2.3 Components for Stem Cell Product Treatment

1. Ex vivo expanded $CD90^+$, $CD105^+$, and CD 73^+ dental pulp-derived stem cells (DPSCs).
2. Sterile PBS.
3. Tissue culture media.
4. Fetal bovine serum.
5. Medium supplements (antibiotics, glutamine).
6. Sterile biosafety cabinet for cell culture.
7. Tissue culture incubator.
8. Tissue culture accessories.
9. Sterile biosafety cabinet in animal vivarium.
10. Centrifuge tubes and holding racks.
11. Aerosol spraying apparatus (airbrush for make-up).

3 Methods

3.1 Induction of Type 1 Diabetes Mellitus (T1DM) in Mice Using Streptozotocin (STZ)

1. House mice for a period no shorter than 1 week in a sterile and Institutional Animal Care and Use Committee (IACUC)-approved animal facility prior to first STZ injection.
2. Allow mice free access to food and water.
3. Weigh all mice accurately and divide them randomly into experimental and control groups.
4. Four hours prior to STZ injection, remove all food from cages for all groups, providing water as normal.
5. Measure blood glucose from a tail-vein blood sample using a blood glucose monitoring system to collect the baseline glucose level of mice using blood glucose meter.
6. Before injection, swipe the surrounding areas with the alcohol swab.
7. Perform intra peritoneal (i.p.) injections of STZ at 40 mg/kg in experimental group of animals. Inject equal volume of citrate buffer i.p. into control group of mice.
8. Place the mice gently back into their home cages (*see* **Notes 6–10**). Provide free access to normal food and water.
9. Repeat **steps 3–6** on days 2 through 5 (the next four consecutive days).
10. On day 14 (9 days after the last STZ injection) allow mice to fast for 4–6 h, then measure blood glucose from a tail-vein blood sample using a blood glucose monitoring system to ensure hyperglycemic conditions in STZ-treated groups. Blood glucose levels >300 mg/dl considered to be T1DM.
11. Perform blood monitoring of glucose levels weekly to ensure hyperglycemia persists.

3.2 DPSC Isolation and Expansion

1. Tissue dissection to be performed in a surgical suite or in a sterile chemical hood (not in tissue culture hood).
2. Discarded third molar teeth are a convenient tissue for extraction of $CD90^+$, $CD105^+$, and $CD73^+$ DPSCs. Place discarded teeth in 15 mL tubes with sterile PBS with antibiotic and antimycotic solutions. Wash exterior of tubes with ethanol prior to placement in tissue culture hood.
3. Using surgical scalpels, cut tissue into ~1 mm piece until tissue resembles a putty-like consistency; it should not separate into pieces but instead become gelatinous in nature, sticking to the blades themselves.
4. Transfer tissue sections onto a tissue culture plate and allow to dry/adhere for ~15 min.

5. Incubate adhered tissues in media supplemented with 20% fetal bovine serum (FBS).
6. Expand in monolayer culture through passage II and perform flowcytometry analysis to verify cell phenotype (*see* **Note 4**).
7. Cells that show to be immunonegative for CD34 and CD45 while being immunopositive for $CD90^+$, $CD105^+$, and $CD73^+$ by flowcytometry will be considered DPSCs.

3.3 Preparation of DPSC Secretory Products

1. DPSC will be cultured in DMEM medium supplemented with 20% FBS and 1% PSG (Penicillin Streptomycin and glutamine) until ~70–80% confluent.
2. Once desired cell confluency is achieved, supplemented media will be discarded and cells will be washed with sterile PBS at least three times.
3. Washed cells are to be cultured in sterile PBS for a desired length in time (e.g., 12 h) at which time the secretory product containing PBS will be collected.
4. Cellular debris is removed by centrifugation at 2000 rpm ($850 \times g$) for 10 min.
5. After centrifugation, supernatants will be collected.
6. Make aliquots of desired volume (e.g., 1 mL/tube). Then transport tubes to animal vivarium facility maintaining sterile conditions at 4 °C.
7. Transport sterile vehicle PBS in separate tubes to the animal facility maintaining sterile conditions at 4 °C.
8. Then in the mouse room, within BSL-2 cabinet, fill aerosol-spraying apparatus with conditioned PBS/vehicle PBS. Maintain aseptic technique.

3.4 Full Thickness Cutaneous Wound Excision and Stem Cell Product Application

1. Anesthetize mouse and remove hair from the lateral dorsal region using a scalpel or hair removal cream.
2. Sanitize area using ethanol.
3. Using a biopsy tissue punch of the desired size (e.g., 5 or 8 mm), excise a full cutaneous wound from the hair removed area. Use of additional surgical equipment may be required (e.g., scalpel, scissor).
4. Using aerosol machine, apply DPSC product/vehicle PBS (*see* **Note 11**) in a manner that is concurrent with experimental design (e.g., two times/day).
5. Place the mouse into the home cage, using care, and observe its movement as it awakens from the anesthesia to ensure its survival and normal activities.
6. Continue with topical application of treatment described in **step 3** for a period of 15–20 days.

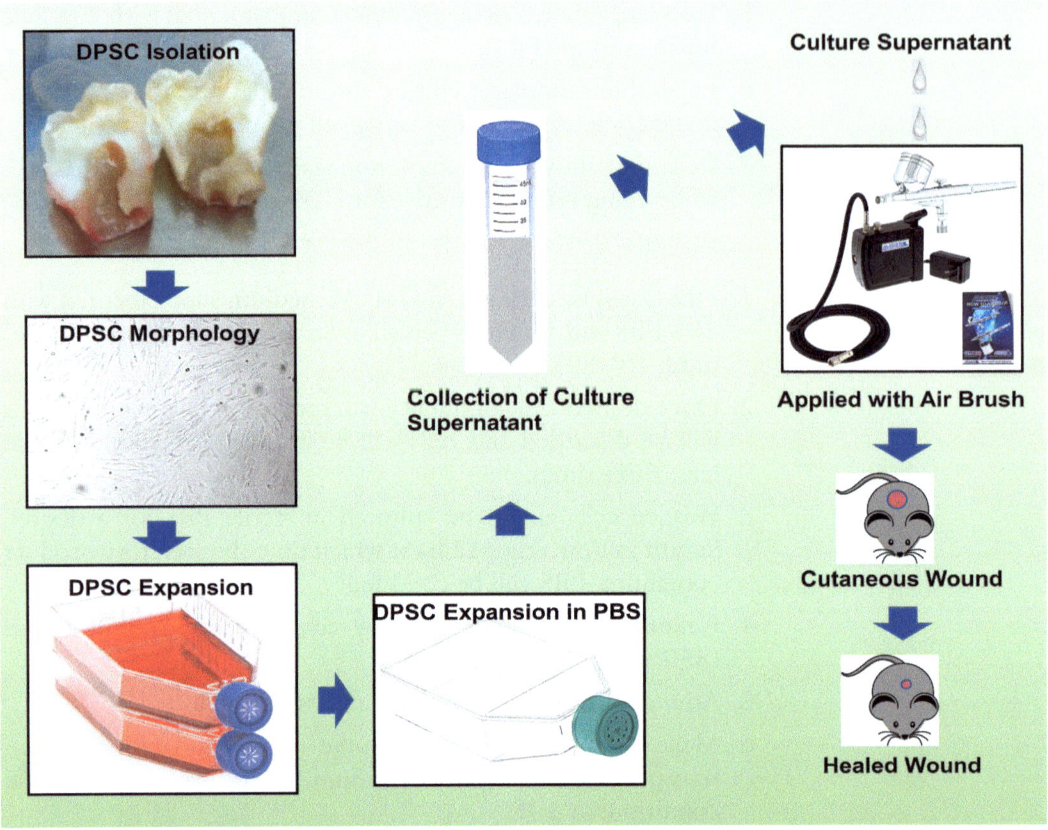

Fig. 1 Schematics showing isolation, expansion, and application of dental pulp–derived stem cell product for cutaneous wound healing in STZ-induced diabetic mice. DPSCs were isolated from human wisdom teeth and were expanded in a cell culture flask. Stem cell secretory products were isolated in PBS and collected. After removing cells and debris, secretory products were applied to the cutaneous wounds using airbrush into a mouse after induction of diabetes with STZ

7. After 15–20 days of topical applications, sacrifice mice and harvest wound bed tissues for the analysis of mRNA, protein expression, and histology (Fig. 1).

4 Notes

1. All the procedures need to be carried out at room temperature and in sterile conditions unless otherwise stated.
2. Always check solutions for contamination with any kind of cellular growth and should be precisely observed against light before each use.
3. Always wear gloves and lab coats when performing any tissue culture experiments.

4. Resuspend cell pellets nicely so that they are devoid of any clumps; it should be single cell suspension.
5. Always wear personal protective equipment (PPE) when working with mice.
6. Monitor mice regularly and provide food and water as often as necessary.
7. Cages should be transported by keeping lid closed, and only opened under a sterile hood to avoid exposure of mice to the non-sterile environment.
8. Blood glucose levels are considered an accurate diagnostic tool for diabetes; there is generally no need to measure blood insulin levels.
9. Ensure that mice are caged in disposable cages and are provided with ample amounts of appropriate absorbent bedding as STZ is metabolized by mice for at least 24 h.
10. Discard biohazard materials appropriately.
11. Pay attention that aerosol spray of DPSC product does not make any droplet. Just make it wet.

Acknowledgments

This work was supported in part by National Institutes of Health grants, R01AR068279 (NIAMS), STTR 1R41EY024217 (NEI), and STTR 1R41AG057242 (NIA).

References

1. Bielefeld KA, Amini-Nik S, Alman BA (2013) Cutaneous wound healing: recruiting developmental pathways for regeneration. Cell Mol Life Sci 70(12):2059–2081. https://doi.org/10.1007/s00018-012-1152-9
2. Elliot S, Wikramanayake TC, Jozic I et al (2018) A modeling conundrum: murine models for cutaneous wound healing. J Investig Dermatol 138(4):736–740. https://doi.org/10.1016/j.jid.2017.12.001
3. Kanji S, Das H (2017) Advances of stem cell therapeutics in cutaneous wound healing and regeneration. Mediat Inflamm 2017:5217967. https://doi.org/10.1155/2017/5217967
4. Kanji S, Das M, Aggarwal R et al (2014) Nanofiber-expanded human umbilical cord blood-derived CD34+ cell therapy accelerates murine cutaneous wound closure by attenuating pro-inflammatory factors and secreting IL-10. Stem Cell Res 12(1):275–288. https://doi.org/10.1016/j.scr.2013.11.005
5. Kanji S, Das M, Aggarwal R et al (2014) Nanofiber-expanded human umbilical cord blood-derived CD34(+) cell therapy accelerates cutaneous wound closure in NOD/SCID mice. J Cell Mol Med 18(4):685–697. https://doi.org/10.1111/jcmm.12217
6. Kanji S, Das M, Joseph M et al (2019) Nanofiber-expanded human CD34(+) cells heal cutaneous wounds in streptozotocin-induced diabetic mice. Sci Rep 9(1):8415. https://doi.org/10.1038/s41598-019-44932-7
7. Heyer K, Herberger K, Protz K et al (2016) Epidemiology of chronic wounds in Germany: analysis of statutory health insurance data. Wound Repair Regen 24(2):434–442. https://doi.org/10.1111/wrr.12387
8. Guest JF, Ayoub N, Mcilwraith T et al (2015) Health economic burden that wounds impose on the National Health Service in the UK. BMJ

Open 5(12):e009283. https://doi.org/10.1136/bmjopen-2015-009283

9. Roger VL (2013) Epidemiology of heart failure. Circ Res 113(6):646–659. https://doi.org/10.1161/Circresaha.113.300268
10. Serjeant GR, Serjeant BE, Mohan JS et al (2005) Leg ulceration in sickle cell disease: medieval medicine in a modern world. Hematol Oncol Clin North Am 19(5):943. https://doi.org/10.1016/j.hoc.2005.08.005
11. Papi M, Papi C (2016) Vasculitic Ulcers. Int J Low Extrem Wounds 15(1):6–16. https://doi.org/10.1177/1534734615621220
12. Maroz N, Simman R (2013) Wound healing in patients with impaired kidney function. J Am Coll Clin Wound Spec 5(1):2–7. https://doi.org/10.1016/j.jccw.2014.05.002
13. Burns J, Pieper B (2000) HIV/AIDS: impact on healing. Ostomy Wound Manage 46 (3):30–40, 42, 44 passim; quiz 48-39
14. Anderson K, Hamm RL (2012) Factors that impair wound healing. J Am Coll Clin Wound Spec 4(4):84–91. https://doi.org/10.1016/j.jccw.2014.03.001
15. Padula WV, Gibbons RD, Pronovost PJ et al (2017) Using clinical data to predict high-cost performance coding issues associated with pressure ulcers: a multilevel cohort model. J Am Med Inform Assoc 24(e1):e95–e102. https://doi.org/10.1093/jamia/ocw118
16. Horn SD, Barrett RS, Fife CE et al (2015) A predictive model for pressure ulcer outcome: the wound healing index. Adv Skin Wound Care 28(12):560–572.; ; quiz 573-564. https://doi.org/10.1097/01.ASW.0000473131.10948.e7
17. Frykberg RG, Zgonis T, Armstrong DG et al (2006) Diabetic foot disorders. A clinical practice guideline (2006 revision). J Foot Ankle Surg 45(5 Suppl):S1–S66. https://doi.org/10.1016/S1067-2516(07)60001-5
18. Rolph DN, Deb M, Kanji S et al (2020) Ferutinin directs dental pulp-derived stem cells towards the osteogenic lineage by epigenetically regulating canonical Wnt signaling. Biochim Biophys Acta Mol basis Dis 1866 (4):165314. https://doi.org/10.1016/j.bbadis.2018.10.032
19. Shin L, Peterson DA (2013) Human mesenchymal stem cell grafts enhance normal and impaired wound healing by recruiting existing endogenous tissue stem/progenitor cells. Stem Cells Transl Med 2(1):33–42. https://doi.org/10.5966/sctm.2012-0041
20. Maxson S, Lopez EA, Yoo D et al (2012) Concise review: role of mesenchymal stem cells in wound repair. Stem Cells Transl Med 1 (2):142–149. https://doi.org/10.5966/sctm.2011-0018

Chapter 4

Musculoskeletal Tissue Engineering Using Fibrous Biomaterials

George Tan, Yingge Zhou, and Dilshan Sooriyaarachchi

Abstract

In tissue engineering, scaffolds should provide the topological and physical cues as native tissues to guide cell adhesion, growth, migration, and differentiation. Fibrous structure is commonly present in human musculoskeletal tissues, including muscles, tendons, ligaments, and cartilage. Biomimetic fibrous scaffolds are thus critical for musculoskeletal tissue engineering. Electrospinning is a versatile technique for fabricating nanofibers from a variety of biomaterials. However, conventional electrospinning can only generate 2D nanofiber mats. Postprocessing methods are often needed to create 3D electrospun nanofiber scaffolds. In this chapter, we present two novel electrospinning-based scaffold fabrication techniques, which can generate 3D nanofiber scaffolds in one-station process: divergence electrospinning and hybrid 3D printing with parallel electrospinning. These techniques can be applied for engineering tissues with aligned fiber structures.

Key words 3D electrospinning, Nanofiber scaffold, Musculoskeletal tissue engineering, Hybrid 3D printing

1 Introduction

Biofabrication is defined as "a process that results in a defined product with biological function" [1]. In the last decade, advanced manufacturing techniques coupled with discovery of novel biocompatible materials have shown great potential in applications of tissue engineering. By integrating biological and engineering technologies, construction of biocompatible architectures has facilitated to meet the need for autologous transplant tissues. It is widely accepted that creation of biomimetic cell environment with micro- and nanoscale topographical features resembling the native tissues is critical for tissue engineering. Major breakthroughs have been made creating complex patterns and integrating bioactive cues [2]. Among various topographical geometries, fibers in micro- and nanoscale have drawn extensive attention because most connective and muscle tissues are in fibrous structure [3–5].

Hiranmoy Das (ed.), *Wound Regeneration: Methods and Protocols*, Methods in Molecular Biology, vol. 2193, https://doi.org/10.1007/978-1-0716-0845-6_4,

Electrospinning, a technique for synthesizing nanoscale fibers [6], has been used for engineering various tissues such as bone [7], nerve [8], cartilage [9, 10], muscle [11], tendon [12], and skin [13]. Recent advances have led to development of complex scaffolds comprised of biomimetic structures to recapitulate the topographic heterogeneity of native tissues. The following sections presents two novel methods of fabricating biomimetic scaffolds for musculoskeletal tissues. The first section presents a rapid fabrication technique for three-dimensional (3D) aligned nanofiber scaffolds for muscle, tendon, and ligament tissues [14]. The approach adopts a symmetrically diverged electric field to induce rapid self-assembly of aligned polymer nanofibers into a centimeter-scale architecture between separately grounded bevels. The scaffolds provided biophysical stimuli to facilitate cell adhesion, proliferation, and morphogenesis in 3D space. The second session presents a hybrid manufacturing technique that combines fused deposition modeling with electrospinning to fabricate biomimetic scaffold for meniscus [10]. The scaffold is composed of 3D-printed high modulus rigid plastic layers with electrospun polycaprolactone (PCL) and collagen (Type 1 & 3) mat in-between. The nanofiber filler will enhance the mechanical properties of the system and provide micro topological cues for cells to organize into a patterned structure.

2 Materials

2.1 Tissue Culture in 3D Aligned Nanofiber Scaffolds

PCL (MW: 80,000) pellet, aluminum foils, polylactic acid (PLA) filament for 3D printing, human fibroblasts (ATCC® MRC-5), Eagle's Minimum Essential Medium (EMEM, ATCC®, Manassas, VA), fetal bovine serum (FBS, ATCC®, Manassas, VA), AlamarBlue (ThermoFisher Scientific, Waltham, MA), 4% formaldehyde, Phalloidin CruzFluor™ 488 Conjugate (Santa Cruz Biotechnology, Dallas, TX), 4′,6-diamidino-2-phenylindole (DAPI, Santa Cruz Biotechnology, Dallas, TX), 70% ethanol solution spray, centrifuge tubes, pipettor and pipette tips, 10 ml beakers, phosphate-buffered saline (PBS).

Prepare PCL solution using high-purity solvents and deionized water. Prepare and store PCL solution at room temperature, do not heat solution. Follow all waste disposal regulations when disposing waste materials.

2.2 Engineered Meniscus by Nanofiber Incorporated Scaffold

Prepare all solutions using deionized (DI) water and/or with phosphate-buffered saline (PBS) at room temperature unless otherwise specified. Adhere to all chemical storage and waste disposal regulations when handling and disposing of materials.

1. Electrospinning Solution: Mix 10% polycaprolactone (PCL, MW = 80,000) pellets and Type-I collagen at 1:1 ratio with hexafluoroisopropanol (HFIP, 99.5%). Place the mixture on a magnetic stirrer and stir at room temperature for 4–6 h.
2. Cell Culture Medium: Mix Eagle's Minimum Essential Medium (EMEM, ATCC® 30-2003™) with 10% fetal bovine serum (FBS, ATCC® 30-2020™) at room temperature (*see* **Note 1**).
3. Phalloidin Fluorescent Stain: Add 1 μl of 1000× Phalloidin stock conjugate in DMSO solution to 1 ml of PBS with 1% bovine serum albumin (BSA) and mix well using the pipette at room temperature.
4. DAPI Fluorescent Stain: Add 2 ml of DI water or dimethylformaldehyde (DMF) to the entire content of the DAPI vial to make 14.3 mM DAPI solution. Add 2.1 μl of this in to 100 μl of PBS to acquire 300 μM DAPI dilution. Further, dilute the solution at 1:1000 ratio to get working 300 nM working DAPI solution.

3 Methods

3.1 Tissue Culture in 3D-Aligned Nanofiber Scaffolds

3.1.1 3D-Printed PLA Collector

1. Design the collector for divergence electrospinning using computer-aided design (CAD) software. The length and width of the collector are 20 mm. The height is 5 mm. The inclination angle for the two bevels is 45°.
2. 3D print the collectors using PLA filament by a fused deposition modeling machine.
3. Cover the inner surfaces of the two bevels with aluminum foil.
4. Ground the foil through wire connectors.
5. Sterilize the collector by ethanol.

3.1.2 PCL Solution

1. Dissolve 15 g of PCL pellets (MW: 80,000) into 100 ml of *N*, *N*-dimethylformamide (DMF) and chloroform (1:1) to make 15% w/v PCL solution.
2. Magnetic stir the solution on a magnetic stirrer for at least 2 h at room temperature until all PCL is dissolved, and the solution is homogeneous.
3. Sterilize the collected nanofibers with UV light in a biosafety cabinet overnight.

3.1.3 Create 3D Nanofiber Scaffold by Divergence Electrospinning

1. Spray 70% ethanol in the electrospinning chamber to fully sterilize electrospinning process.
2. Turn on the electrospinning machine and load PCL solution into syringe.

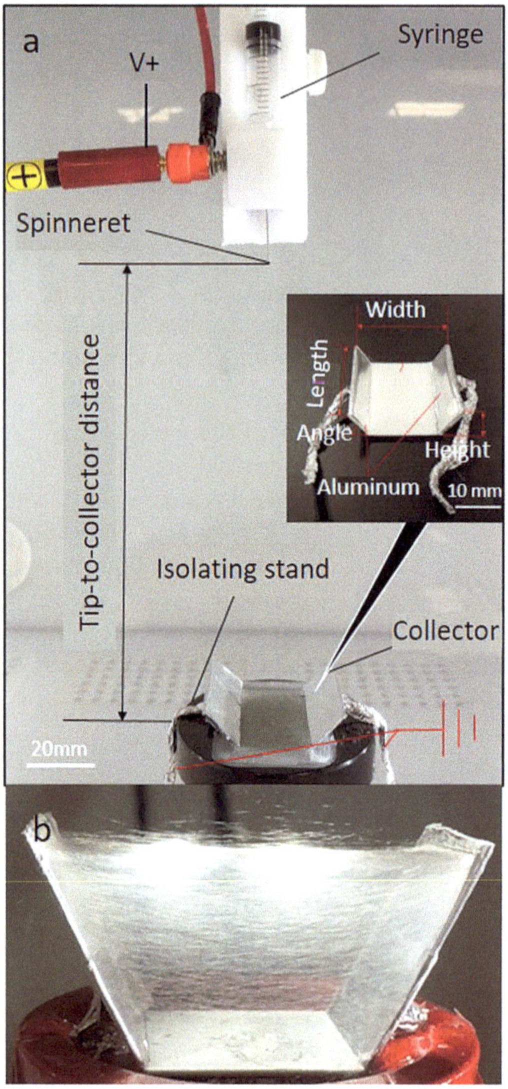

Fig. 1 (**a**) Configuration of divergence electrospinning, (**b**) 3D-aligned nanofiber scaffold

3. Set syringe and collector in position, collector should be below the needle tip, the tip-to-base distance was set to 90 mm, close the door (Fig. 1). Set the applied voltage as 10 kV. The nozzle size for syringe is gauge 22.
4. Turn on the micropump and set the pump rate to 0.5–1 ml/h. Wait until solution comes out from needle tip in a constant and steady way.
5. Apply preset voltage. Adjust the pump rate and voltage in real time if necessary to maintain a stable electrospinning process. Clean the needle tip during the process if the solution accumulates at the tip. Keep electrospinning for 3 min (*see* **Note 5**).

6. Sterilize the collected nanofibers with UV light in a biosafety cabinet overnight (*see* **Note 2**).
7. Sterilize a 10 ml beaker with ethanol, and position the collector with nanofibers into the beaker vertically, until all the collector was inside the beaker. Perform the sterilization in the biosafety cabinet. Air-dry the beaker for overnight.

3.1.4 Cell Culture

1. Prepare 10 ml of complete EMEM with 10% FBS.
2. Thaw the fibroblast cells. Centrifuge the cells at 200 × *g* for 5 min and resuspend the cells in fresh complete medium.
3. Deliver the cells to the 3D nanofiber scaffold by slowly dropping the medium into the beaker using a pipettor. Make one drop at a time. Drop the cells onto fibers as more as you can. Cover the beaker top with sterilized parafilm and move the beaker into the incubator.
4. Change the medium after 4 h. Do not break the nanofibers. Observe cell morphology under the microscope. Continue culturing the cells in the incubator at 37 °C and 5% CO_2.
5. Conduct the alamar blue test as per the supplier's protocol at days 2, 7, and 14. Remove the residual medium and transfer the whole collector with nanofibers into another sterilized 10 ml beaker. This is to eliminate the effect of cells attached to the bottom of the beaker while just keeping the cells on scaffold. Reload the new beaker with new complete medium. Conduct this move in biosafety cabinet (*see* **Note 3**).
6. Change the medium every 2 days and take microscopic images every day.
7. After day 14, fix the cells with 4% formaldehyde and stain cells with Phalloidin CruzFluor™ 488 Conjugate and DAPI for filamentous action and cell nucleus, respectively, as per suppliers' protocols. Take fluorescence images from the top and side view of the scaffold under microscope (Fig. 1) (*see* **Note 4**, **6**, and **7**).

3.2 Engineered Meniscus by Nanofiber Incorporated Scaffold

3.2.1 Preparation of 3D-Printed Meniscus Via Multijet Printing (MJP)

1. Prepare (.stl) file for the meniscus using a CAD software. Use 3D print slicer software to prepare the file for printing via ProJet MJP 2500 system (3D Systems, Inc., Rock Hill, SC).
2. Once the print is complete, move the print platform into the freezer unit and leave it in for 10–15 min. Remove the printed parts and move them into the MJP EasyClean System (3D Systems, Inc., Rock Hill, SC).
3. Place the parts in the strainer basket and put the strainer in the Bulk Wax removal system, with the bulk wax system heated to 10. Close cover and leave in the bulk wax system for approximately 15–20 min after steam is being generated.

4. Wearing insulating gloves, use the strainer basket removal handles to pick the basket up. With the strainer still over the wax catch pan, tilt it to the corner to allow liquid wax to drain away. Now place the strainer basket with your printed parts in the fine wax removal system and leave it in for 5–10 min.
5. Wash the printed parts thoroughly with warm soapy water and then with isopropyl alcohol.

3.2.2 Electrospinning

1. Prepare the electrospinning chamber and set the electrospinning parameters.
   (a) Tip to collector distance—12.5 cm.
   (b) Distance between the metal bars—3 cm.
   (c) Voltage at needle—12 kV.
   (d) Metal bars connected to the negative potential—(−) 0.1 kV.
   (e) Pump rate—2 ml/h.
2. Continue electrospinning for 5 min and collect highly aligned nanofibers between the plates. Once complete, switch power off and remove the bars along with the platform.
3. Wet the 3D-printed meniscus scaffold with isopropyl alcohol. Place the part in between the metal bars and scoop the fibers onto the scaffold. Immediately place it inside a sterilized petri dish and close the lid.
4. Move the 3D-printed part with highly electrospun nanofibers into the biosafety cabinet. Place them inside the 12-well cell culture plate.
5. Attach the scaffolds on to the bottom of the well using a 15% PCL solution to prevent them from floating in the cell medium.
6. Sterilize the whole system under UV light for 90 min. Now they are ready for cell seeding.

3.2.3 Cell Culture

1. Remove the frozen human osteosarcoma (MG-63, ATCCP®P CRL-1427™) cells from the freezer and thaw them at 37 °C in a water bath.
2. Remove the contents in the vial into a centrifuge tube and centrifuge for 5 min at 200 × *g*.
3. Carefully remove the medium without disturbing the sedimented cells at the bottom.
4. Carefully wash the vial with 1 ml of PBS to remove any residue cell medium.
5. Add 1 ml of the medium to the tube and mix the cells.

6. To count the number of cells, remove 100 µl of the mixed medium and add 100 µl of typhon blue.
7. Add 10 µl of the solution into the hemocytometer and count the number of cells.
8. Dilute the solution in the centrifuge tube, so an approximately 100,000 cells can be added to each well.
9. Carefully wash each well on the culture plate with PBS three times.
10. Using a micropipette, add 3 ml of the cell solution into each well.
11. Use a microscope to ensure that cells are trapped between the nanofibers.
12. Spray the outside of the culture plate with isopropyl alcohol to sterilize. Then move it into the incubator at 37 °C with 5% CO_2.
13. After 24 h, move the plate into biosafety cabinet.
14. Carefully remove the medium and wash with PBS three times.
15. Add 3 ml of freshly prepared cell medium into each well.
16. Repeat **steps 14–16** every 3 days.

3.2.4 Alamar Blue Assay

1. Remove all the medium from the well and move the scaffolds to a new cell culture plate. This is done to avoid results being affected by the cells attached to the bottom of the well.
2. Wash with PBS three times.
3. Add new cell medium.
4. Add 10% of alamar blue assay (Thermo Fisher Scientific, Waltham, MA).
5. Incubate the plate for 2 h at 37 °C with 5% CO_2.
6. Measure the absorbance signal by accuSkan™ FC Filter-Based Microplate Photometer (Thermo Fisher Scientific, Waltham, MA).
7. Calculate the percent reduction (Fig. 2).

3.2.5 Fluorescent Imaging

1. Wash the cells with PBS three times.
2. Fix cells with 4% formaldehyde for 15 min. Prepare the 4% formaldehyde solution by diluting the stock solution with PBS (*see* **Note 8**).
3. Wash cells with PBS three times.
4. Treat cells with 0.1% triton X-100 for 10 min (*see* **Note 9**).
5. Wash the cells with PBS three times.

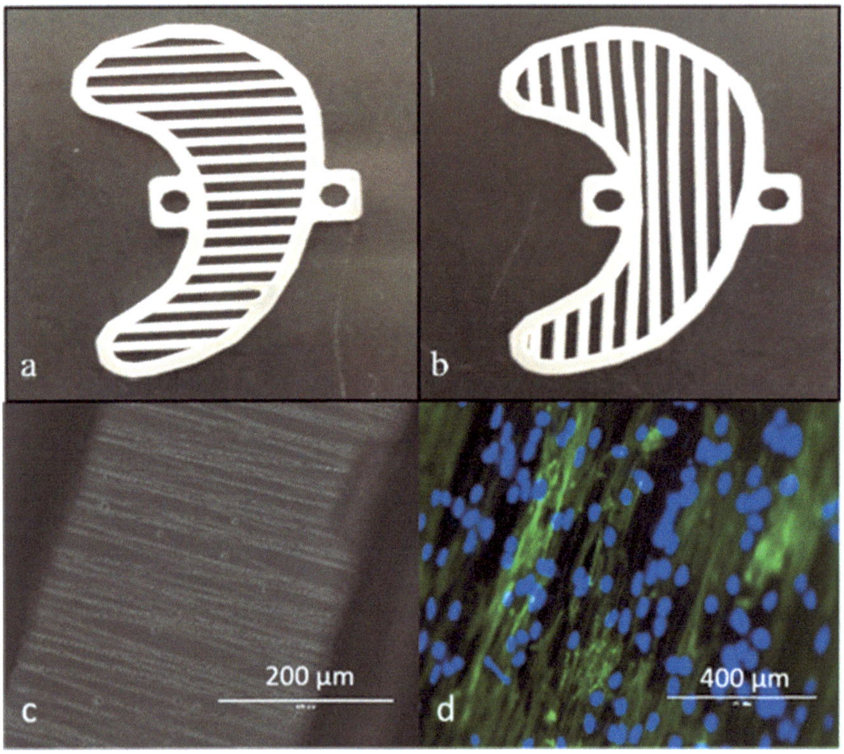

Fig. 2 3D-printed meniscus frames with (**a**) radial lines and (**b**) circumferential lines. (**c**) Aligned nanofibers between the frame lines. (**d**) Aligned cells on the nanofibers

6. Add previously prepared phalloidin working solution enough to cover the surface of the scaffold and incubate for 30 min at room temperature in the dark.
7. Wash cells three times with PBS.
8. Stain the nuclei with previously prepared DAPI solution and incubate at room temperature for 5 min in the dark.
9. Wash the cells with PBS three times.
10. Observe the cells using the fluorescent microscope.

3.2.6 Compression Test

1. Stack 12 scaffolds on top of each other. Ensure that they are properly aligned.
2. Place them on the bed of Shimadzu Tensile tester (Shimadzu Corporation, model AGS—50k NXD).
3. Conduct the test at 2.5 mm/min with full scale of 50,000 N.
4. Plot the stress—strain curve using the acquired data.
5. Calculate the compression modulus.

4 Notes

1. Conduct all cell-related experiments in the biosafety cabinet.
2. Always sterilize with ethanol before placing anything into biosafety cabinet.
3. Be gentle and careful with the scaffold as the nanofibers can be fragile.
4. Be familiar with the alarmarBlue test, Phalloidin CruzFluor™ 488 Conjugate, and DAPI protocols before conducting the test and staining experiments.
5. Spray with ethanol in the electrospinning chamber to eliminate more contaminated nanofibers as possible. Cover the collected fibers with sterilized container into the biosafety cabinet as soon as possible.
6. DAPI has poor solubility in water. Use a sonicator to disperse the DAPI if necessary.
7. DAPI solution may be stored up to 6 months or longer at $\leq 20^{\circ}$C.
8. Formaldehyde fixes tissue by crosslinking the proteins. Fixation ensures following four key aspects.
   (a) It preserves and stabilizes cell morphology and tissue architecture.
   (b) It inactivates proteolytic enzymes that could otherwise degrade the sample.
   (c) It strengthens samples so that they can withstand further processing and staining.
   (d) It protects samples against microbial contamination and possible decomposition.
9. Triton X 100 is a detergent that permeabilizes the cell membrane.

References

1. Groll J, Boland T, Blunk T et al (2016) Biofabrication: reappraising the definition of an evolving field. Biofabrication 8(1):013001
2. Jin G, He R, Sha B et al (2018) Electrospun three-dimensional aligned nanofibrous scaffolds for tissue engineering. Mater Sci Eng C Mater Biol Appl 92:995–1005
3. Tan GZ, Zhou Y (2019) Electrospinning of biomimetic fibrous scaffolds for tissue engineering: a review. Int J Polym Mater Polym Biomater:1–14
4. Zhou Y, Hu Z, Du D et al (2019) The effects of collector geometry on the internal structure of the 3D nanofiber scaffold fabricated by divergent electrospinning. Int J Adv Manuf Technol 100(9–12):3045–3054
5. Zhou Y, Mahurubin S, Sooriyaarachchi D et al (2019) The effect of nanoclays on nanofiber density gradient in 3D scaffolds fabricated by divergence electrospinning. Procedia Manufacturing 34:110–117
6. Zhou Y, Thakurathi M, Quitevis EL et al (2019) Electrospinning 3D nanofiber structure of polycaprolactone incorporated with silver nanoparticles. JOM 71(3):956–962

7. Shao W, He J, Sang F et al (2016) Coaxial electrospun aligned tussah silk fibroin nanostructured fiber scaffolds embedded with hydroxyapatite–tussah silk fibroin nanoparticles for bone tissue engineering. Mater Sci Eng C 58:342–351
8. Hu J, Kai D, Ye H et al (2017) Electrospinning of poly (glycerol sebacate)-based nanofibers for nerve tissue engineering. Mater Sci Eng C 70:1089–1094
9. Zhou Y, Chyu J, Zumwalt M (2018) Recent Progress of fabrication of cell scaffold by electrospinning technique for articular cartilage tissue engineering. Int J Biomater 2018:1953636
10. Sooriyaarachchi D, Wu J, Feng A et al (2019) Hybrid fabrication of biomimetic meniscus scaffold by 3D printing and parallel electrospinning. Procedia Manufacturing 34:528–534
11. Laurent C, Liu X, De Isla N et al (2018) Defining a scaffold for ligament tissue engineering: what has been done, and what still needs to be done. J Cell Immunother 1:4–9
12. Nowlin J, Bismi MA, Delpech B et al (2018) Engineering the hard–soft tissue interface with random-to-aligned nanofiber scaffolds. Nano 5:1–9
13. Vázquez N, Chaires C, Hernández B et al. (2016) Poly (lactic-co-glycolic acid)/gelatin electrospun scaffolds seeded with human mesenchymal stem cells for skin tissue engineering. In: Front. Bioeng. Biotechnol. Conference Abstract: 10th World Biomaterials Congress. doi: https://doi.org/10.3389/conf. FBIOE
14. Tan GZ, Zhou Y (2018) Tunable 3D nanofiber architecture of polycaprolactone by divergence electrospinning for potential tissue engineering applications. Nanomicro Lett 10(4):73

Chapter 5

Cutaneous Wound Generation in Diabetic NOD/SCID Mice and the Use of Nanofiber-Expanded Hematopoietic Stem Cell Therapy

Sarah Anderson and Hiranmoy Das

Abstract

Despite significant advances in diabetic wound management, diabetic wounds remain a significant global problem that decreases patient's quality of life, and chronic wounds may lead to amputation and death to the patients. To develop a potential regenerative therapy, a xenogeneic transplantation compatible laboratory model needs to be developed. This procedure demonstrates how to isolate hematopoietic stem cells ($CD133^+$) from human umbilical cord blood, expand $CD34^+$ stem cells using a nanofiber scaffold (polyether sulfone-coated and amino group-treated), induce diabetes in immunocompromised (NOD/SCID) mice, induce a cutaneous wound in mice, and how to treat the wound with the nanofiber-expanded $CD34^+$ stem cells. This protocol also shows how to measure wound healing.

Key words Diabetes induction, Cutaneous wound, NOD/SCID mice, $CD34^+$ stem cell

1 Introduction

Despite significant advances in diabetic wound management, diabetic wounds remain a significant global issue with a prevalence of 6.3% among diabetic patients. In the United States, this problem is even more pronounced with a prevalence of 13% [1]. Patients with diabetes are at risk of developing chronic wounds such as pressure ulcers and foot ulcers [2]. These ulcers can be a major source of infection that significantly decreases patients' quality of life. Patients with diabetes have impaired healing for a variety of reasons including but not limited to impaired inflammatory cell function, impaired growth factor production, impaired fibroblast proliferation, and migration [3]. Current treatment for diabetic wounds involves debridement, dressing the wound, relieving pressure from the wound, controlling infection, assessing the vasculature, and controlling blood sugar [4]. However, even with these treatments, diabetic foot ulcers precede 85% of nontraumatic lower-limb

Hiranmoy Das (ed.), *Wound Regeneration: Methods and Protocols*, Methods in Molecular Biology, vol. 2193,
https://doi.org/10.1007/978-1-0716-0845-6_5, © Springer Science+Business Media, LLC, part of Springer Nature 2021

amputations [5]. There have been several preclinical studies looking at the efficacy of stem cell treatment as a potential regenerative therapy for healing diabetic ulcers. Preclinical studies showed that stem cell therapy decreased inflammation, increased neovascularization, increased infiltration of fibroblasts, and restored the extracellular matrices [6–8]. There was recently a phase 2 clinical trial evaluating the treatment of ALLO-ASC-DFU, a hydrogel containing allogeneic adipose-derived mesenchymal stem cells, for the treatment of diabetic foot ulcers. The study demonstrated a statistically significant reduction in wound size compared to the control [9]. These studies highlight the importance of preclinical data so that more therapies can be developed.

2 Materials

These experiments must be performed in approved BSL2 laboratories (*see* **Note 1**). Procedures should be prepared at room temperature in a sterile condition (*see* **Note 2**). When working with human fluids, Institutional Review Board approval must be obtained first. Institutional Animal Care and Use Committee must approve all work using animals before initiation of experiments. Dispose of waste in appropriate receptacles whether it is a trashcan, biohazard waste container, or sharps container (*see* **Notes 3** and **4**).

2.1 Isolation and Expansion of $CD34^+$ Stem Cells

1. Human umbilical cord blood.
2. "AutoMACS" cell sorting machine.
3. Ficoll-Paque.
4. Flow cytometer.
5. Nanofiber scaffolds coated 24-well cell culture plate [10, 11].
6. StemSpan serum-free expansion media (SFEM) [12].
7. BSL2 biosafety cabinet.
8. Cell culture incubator.
9. Tabletop centrifuge.
10. Water bath.
11. Ice bucket.
12. 4 °C refrigerator/incubator.

2.2 Induction of Diabetes

1. NOD/ SCID male mice 8–10 weeks old; 6–9 mice/group.
2. Streptozotocin (STZ) in powder form should be kept from the light and stored in 2–8 °C. Doses of 50 mg/kg should be made in citrate buffer (pH—4.2) immediately before injections. No more than 3 ml of a solution should be made for each i.p. injection (*see* **Notes 6** and **8**).

3. BSL2 biosafety cabinet in the animal facility.
4. Anesthesia unit (*see* **Note 9**).
5. 25- to 27-gauge needle for intraperitoneal injections.
6. Lancet to nick the tail to measure glucose in the blood.
7. Alphatrak blood glucose monitoring system.

2.3 Induction of Cutaneous Wound

1. Induced diabetic NOD/SCID mice.
2. Razor.
3. Betadine solution.
4. 8 mm punch.

2.4 Stem Cell Treatment

1. Ex vivo nanofiber-expanded $CD34^+$ stem cells from human umbilical cord blood.
2. Serum-free media as a vehicle.
3. Syringes and a 25-gauge needle, or smaller, for tail vein injection.
4. Diabetic wounded mice.
5. BSL-2 hood in the animal facility.

3 Methods

3.1 Isolation and Expansion of $CD34^+$ Stem Cells

1. Obtain human umbilical cord blood with prior IRB approval.
2. Isolate mononuclear cells by Ficoll separation (*see* **Note 5**).
3. Use an AutoMACS to separate $CD133^+$ cells (primitive stem cell marker) and confirm the purity of population by flow cytometry.
4. Seed 800 $CD133^+$ cells on each well of a 24-well cell culture plate, which are containing glued aminated polyethersulfone nanofiber scaffold in 600 μl of StemSpan SFEM with essential supplements.
   (a) 1% BSA.
   (b) 0.01 mg/ml recombinant human insulin.
   (c) 0.2 mg/ml human transferrin.
   (d) 0.1 mM 2-mercaptoethanol.
   (e) 2 mM L-glutamine in Iscove's MDM.
   (f) 0.04 mg/ml low-density lipoprotein (Athens Research and Technology Inc., USA).
   (g) 100 ng/ml purified recombinant stem cell factor (SCF) (Peprotech Inc., Rocky Hill, NJ, USA).
   (h) 100 ng/ml Flt-3 ligand (FLT3) (Peprotech Inc., Rocky Hill, NJ, USA).

(i) 50 ng/ml TPO (Peprotech Inc., Rocky Hill, NJ, USA).

(j) 20 ng/ml IL-3 (Peprotech Inc., Rocky Hill, NJ, USA).

5. Culture the cells for 10 days at 37 °C in 5% CO_2 without changing medium; this will yield a large population of $CD34^+$ stem cells.

3.2 Induction of Diabetes in Mice

1. Collect a drop of blood from the tail to measure baseline blood glucose level (*see* **Note 10**).
2. Fast mice for 4 h before STZ injection. Inject 50 mg/kg of STZ intraperitoneally in the NOD/SCID mice for 5 consecutive days (*see* **Notes 7** and **8**).
3. Check for the induction of diabetes by measuring blood glucose levels after 7 days of the first injection. Briefly anesthetize the mice and nick the tail with a lancet (*see* **Notes 9** and **10**). Place a drop of blood on the blood glucose monitoring system. Mice who have blood glucose higher than 300 mg/dl are considered diabetic in the NOD/SCID mouse model. The blood glucose is measured once a week for 4 weeks to ensure the induction of diabetes and stability.

3.3 Induction of Cutaneous Wound

1. Four weeks after the STZ injection, prepare the mouse for a punch biopsy. Under anesthesia, shave the fur on the dorsal region and wipe the skin with betadine, an antiseptic solution (*see* **Note 9**).
2. Using a full-thickness 8-mm punch creates a wound in the prepared dorsal area of the mouse.

3.4 Preparation of Stem Cells

1. Collect $CD34^+$ nanofiber-expanded human umbilical cord blood-derived stem cells in the BSL-2 biosafety cabinet.
2. Count cells under a microscope using a hemocytometer/automated cell counter.
3. Centrifuge the cells for 5 min at 1400 rpm ($416 \times g$), discard media, and break the pellet.
4. Resuspend the cells in serum-free media. Calculate the amount of media needed using the ratio of 500,000 cells/200 μl of volume for a single mouse.
5. Make aliquots of cells and serum-free media (1 ml/tube) in microcentrifuge tubes. Transport the tubes to the vivarium in a sterile condition at 4 °C.
6. In the mouse room, in a BSL-2 cabinet, fill the syringe with 200 μl of suspended cells or serum-free media as a vehicle.

3.5 Treat Mice with Stem Cells

1. Two hours after cutaneous excision, inject the nanofiber-expanded CD 34^+ stem cells (0.5×10^6 cells/mouse) into the lateral tail vein of the mouse (*see* **Notes 8**, **10**, and **11**) (Fig. 1).

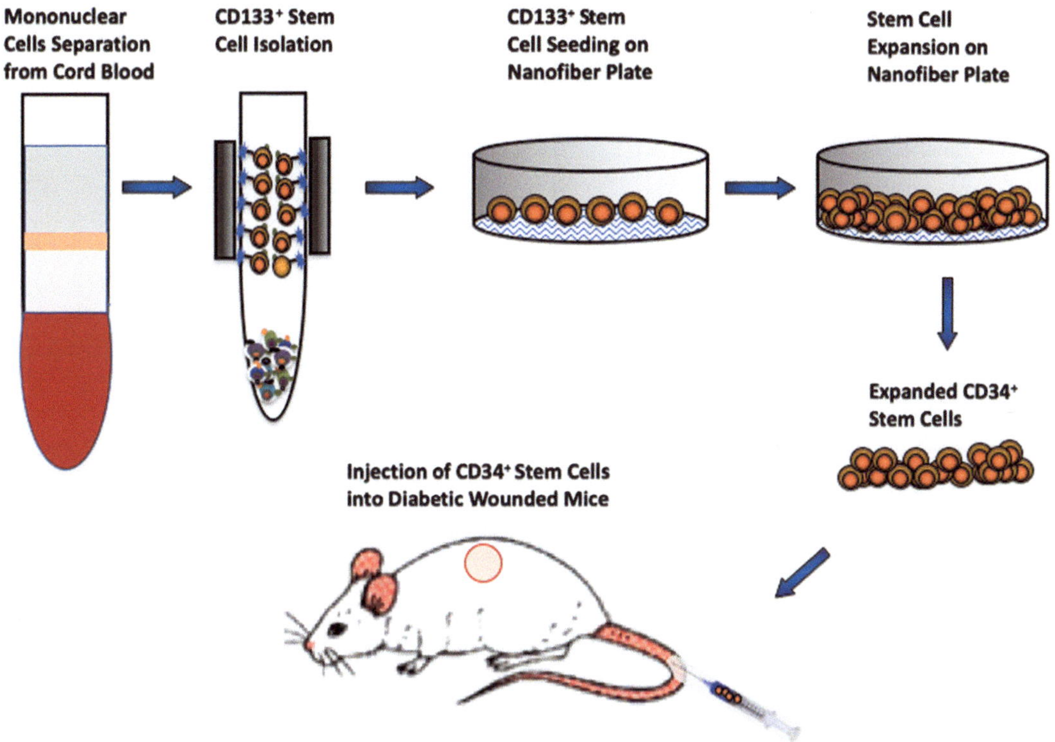

Fig. 1 Schematics showing isolation, expansion, and injection of hematopoietic stem cells. Mononuclear cells were isolated from human umbilical cord blood using Ficoll-Paque centrifugation method; AutoMACS-separated $CD133^+$ (primitive phenotype) stem cells were seeded and expanded on nanofiber scaffolds. $CD34^+$ (mature phenotype) stem cells were collected and injected through the lateral tail veins of wounded diabetic mice

   (a) The stem cells are in 200 μl of serum-free media.
   (b) Sham mice only receive the serum-free media.
2. Sacrifice the mice on the 5th, 10th, and 15th days after wounding/stem cell therapy.
3. Take a skin sample of the wound and 2 mm of tissue around the wound at each time point.

3.6 Evaluate Wound Healing

1. Take a picture of the wound every other day from day 0 to day 15 at a fixed distance. Day 0 is considered the control (Fig. 2).
2. Measure the wound by tracing the wound area onto acetate paper.
3. Digitize the tracings and calculate the wound area by using the University of Texas Health Science Center at San Antonio image tool version 3.00, and calculate the percent of wound closure.

   The formula used to calculate the percent of healing is as follows:

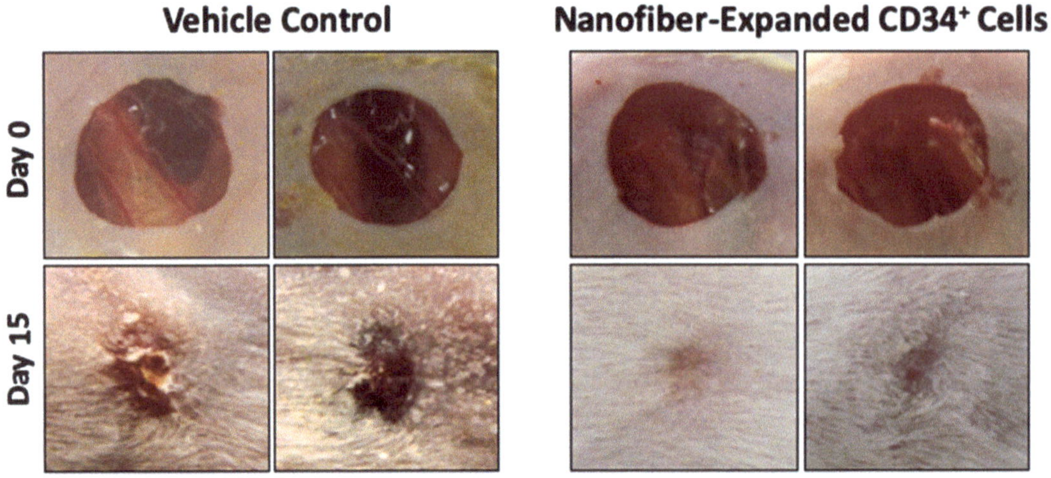

Fig. 2 Representative images showing the effect of stem cell therapy for diabetic wound healing. Diabetes was induced by STZ injection in both groups of mice, and after a month of stabilization, wounds were generated and considered day 0. Stem cells or vehicle was injected into the lateral tail veins of the wounded diabetic mice 2 h after the excision. By day 15, vehicle control mice did not heal completely and remained with open wounds; however, mice received $CD34^+$ stem cells were healed. (Images adapted from Kanji, S., et al. *Sci Rep*, 2019. **9**(1): p. 8415)

$$\frac{\text{area of wound on day } 0 - \text{area of wound on day of examination}}{\text{area of wound on day } 0} \times 100$$

3.7 Evaluate Wound Healing Mechanistically

1. Harvest skin tissues after sacrificing mice.
2. Perform immunohistochemistry and stain for von-Willebrand Factor (vWF) and myeloperoxidase (MPO).
3. Isolate mRNA from the wound tissues and perform quantitative RT-PCR analysis. Analyze the relative expression of the genes Mouse (m) IL-1β, m TNF-α, m IL-10, and m β-actin. Normalize the mRNA expression levels in β-actin.
4. Isolate protein from wound edge tissue and perform Western blot analysis. Use the antibodies MMP-1, MMP-2, p65, and GAPDH. Normalize the protein expression with GAPDH.
5. Perform a total collagen assay from tissue samples; use the Sircol Collagen Assay kit (*see* **Note 12**).

4 Notes

1. Must be trained to work in a BSL2 laboratory using human blood.
2. Unless otherwise stated, make all reagents and carry out all procedures in a sterile condition at room temperature.
3. Check all solutions and reagents for contamination before use.

4. Always wear appropriate personal protective equipment (PPE) when working with tissue culture experiments and when working with mice.
5. See isolation of mononuclear cells methodology and applications by GE Healthcare and Life Sciences for manufacturer's protocol on Ficoll-Paque separation.
6. Streptozotocin should be stored in a cabinet at the animal facility to avoid contamination during transport from laboratory to animal facility.
7. Intraperitoneal (i.p.) injections should be performed as follows: using a one-handed hold tip the mouse so that its peritoneum is facing upwards and its head is tipped toward the ground. Keep the needle at a 30° angle when injecting into the abdomen. Inject into the lower right quadrant just above the hip.
8. For all injections, ensure that there are no bubbles in the syringe.
9. Many different agents can be used for anesthesia. Regardless of method, it is very important to monitor the mice to ensure that there is successful anesthesia, as well as avoiding depression of cardiac and respiratory functions.
10. Place the mouse in a tube restrainer to collect blood from the tail and inject the cell suspension into the lateral tail vein.
11. For lateral tail vein injections, restrain the mouse and dilate the lateral tail vein. This can be done by using warm water to dilate the vein.
12. Sircol "Collagen Assay kit" provides the protocol for performing a total collagen assay.

References

1. Zhang P, Lu J, Jing Y et al (2017) Global epidemiology of diabetic foot ulceration: a systematic review and meta-analysis (dagger). Ann Med 49(2):106–116. https://doi.org/10.1080/07853890.2016.1231932
2. Sen CK, Gordillo GM, Roy S et al (2009) Human skin wounds: a major and snowballing threat to public health and the economy. Wound Repair Regen 17(6):763–771. https://doi.org/10.1111/j.1524-475X.2009.00543.x
3. Patel S, Srivastava S, Singh MR et al (2019) Mechanistic insight into diabetic wounds: pathogenesis, molecular targets and treatment strategies to pace wound healing. Biomed Pharmacother 112:108615. https://doi.org/10.1016/j.biopha.2019.108615
4. Everett E, Mathioudakis N (2018) Update on management of diabetic foot ulcers. Ann N Y Acad Sci 1411(1):153–165. https://doi.org/10.1111/nyas.13569
5. Alavi A, Sibbald RG, Mayer D et al (2014) Diabetic foot ulcers: part I. pathophysiology and prevention. J Am Acad Dermatol 70(1): e1–e18.; ; quiz 19-20. https://doi.org/10.1016/j.jaad.2013.06.055
6. Kanji S, Das M, Joseph M et al (2019) Nanofiber-expanded human CD34(+) cells heal cutaneous wounds in streptozotocin-induced diabetic mice. Sci Rep 9(1):8415. https://doi.org/10.1038/s41598-019-44932-7
7. Chen CY, Rao SS, Ren L et al (2018) Exosomal DMBT1 from human urine-derived stem cells

facilitates diabetic wound repair by promoting angiogenesis. Theranostics 8(6):1607–1623. https://doi.org/10.7150/thno.22958

8. Xu J, Zgheib C, Hodges MM et al (2017) Mesenchymal stem cells correct impaired diabetic wound healing by decreasing ECM proteolysis. Physiol Genomics 49(10):541–548. https://doi.org/10.1152/physiolgenomics.00090.2016
9. Moon KC, Suh HS, Kim KB et al (2019) Potential of allogeneic adipose-derived stem cell-hydrogel complex for treating diabetic foot ulcers. Diabetes 68(4):837–846. https://doi.org/10.2337/db18-0699
10. Joseph M, Das M, Kanji S et al (2014) Retention of stemness and vasculogenic potential of human umbilical cord blood stem cells after repeated expansions on PES-nanofiber matrices. Biomaterials 35(30):8566–8575. https://doi.org/10.1016/j.biomaterials.2014.06.037
11. Chua KN, Chai C, Lee PC et al (2006) Surface-aminated electrospun nanofibers enhance adhesion and expansion of human umbilical cord blood hematopoietic stem/progenitor cells. Biomaterials 27(36):6043–6051. https://doi.org/10.1016/j.biomaterials.2006.06.017
12. Das H, Abdulhameed N, Joseph M et al (2009) Ex vivo nanofiber expansion and genetic modification of human cord blood-derived progenitor/stem cells enhances vasculogenesis. Cell Transplant 18(3):305–318. https://doi.org/10.3727/096368909788534870

Chapter 6

Contusion Rodent Model of Traumatic Brain Injury: Controlled Cortical Impact

Marie-Line Fournier, Tifenn Clément, Justine Aussudre, Nikolaus Plesnila, André Obenaus, and Jérôme Badaut

Abstract

Traumatic brain injury (TBI) is a heterogeneous brain injury which represents one of the leading causes of mortality and disability worldwide. Rodent TBI models are helpful to examine the cellular and molecular mechanisms after injury. Controlled cortical impact (CCI) is one of the most commonly used TBI models in rats and mice, based on its consistency of injury and ease of implementation. Here, we describe a CCI protocol to induce a moderate contusion to the somatosensory motor cortex. We provide additional protocols for monitoring animals after CCI induction.

Key words Traumatic brain injury, Pediatric, Behavior, MRI, Immunohistochemistry

1 Introduction

1.1 Clinical Aspects of Traumatic Brain Injury (TBI)

Traumatic brain injury (TBI) represents one of the leading causes of mortality and disability worldwide, with only for the United States (US) an annual incidence of 1.7 million TBI patients between 2002 and 2006 [1]. It is now estimated that 3.2–5.3 million persons are currently suffering from TBI-related disability, which represents between 1.1% and 1.7% of the total US population [2, 3]. It is well established that TBI, even of mild severity, has long-term complications impacting the daily quality of life [2, 3]. Recently, it has been recognized that TBI represents a complex chronic brain disease, sharing common injury pathways with chronic neurodegenerative disorders [4]. Very importantly, the pediatric population is more vulnerable due to ongoing brain development and, in particular, young children, who are being at the greatest risk to have a TBI (<5 years old). Adolescents (15–19 years old) are the second group affected, followed by older adults (>75 years old) [1]. TBI is frequently caused by a sudden and unexpected external and physical insult to the brain. The variety of external mechanical

Hiranmoy Das (ed.), *Wound Regeneration: Methods and Protocols*, Methods in Molecular Biology, vol. 2193, https://doi.org/10.1007/978-1-0716-0845-6_6,

forces, for example, direct impact to the skull, rapid acceleration and deceleration, blast waves, or penetration injury, has for direct consequence that TBI is very heterogeneous. Clinically, TBI severity has been classified as mild, moderate and severe, using the patient's level of consciousness measured by the Glasgow Coma Scale (GCS) combined with a morphological evaluation of the brain using computerized tomography (CT) [5]. However, these severity categories are not fully standardized and may not accurately reflect all forms of TBI, especially diffuse axonal injury which is not visible on CT. Heterogeneity in TBI is very likely associated with various molecular and cellular mechanisms that can be studied by modeling the injury in rodents [6]. In the literature, there is a large array of different rodent TBI models as summarized below.

1.2 Various Preclinical Models to Study Pathophysiological Mechanisms

The purpose of this chapter is to detail the methods underlying one of the most commonly used rodent brain contusion models, controlled cortical impact (CCI). As an introduction, we briefly review the most common rodent models used to study the molecular and cellular mechanisms after TBI. The primary injury in TBI is produced with a mechanical force directly applied to the brain with various apparatus that have quantifiable amplitude, duration, velocity, and acceleration. In addition, the rodent's head can be fixed or allowed free movement, and with the skull opened or closed (for concussion studies). The three common models of TBI are weight drop (WD), fluid percussion injury (FPI), and controlled cortical impact (CCI) (Fig. 1). Depending of the pathophysiology, the animal models can be classified into five major groups of injury reflecting the clinical symptoms: focal, mixed, diffuse, complex, and other [7]. Based on the classification, the models cited belong to the subgroups focal for CCI, mixed for the WD and lateral FPI, and diffuse for the central FPI (Fig. 1).

The WD model uses a calibrated weight falling from a designated height directly on the skull or on a steel disc glued onto the skull of the rodent to induce TBI. The steel disc helps to spread the mechanical energy of the weight over the skull in order to prevent fractures while still injuring the brain (Fig. 1a). The head of the rodent is not often fixed and placed either on solid surface or on foam with a defined elasticity to allow acceleration and deceleration of the brain. The WD model reproduces many characteristics of closed-head injury (CHI) in humans. It is a mix between focal and diffuse injuries with wide variety of severities from mild to severe. For example, the WD model developed by Dr. Marmarou' group reproduces several features of diffuse axonal injury (DAI) observed in TBI patients, such as diffuse perivascular edema, small intraparenchymal hemorrhages, widespread axonal injury, and increase of intracranial pressure (ICP) [8, 9].

FPI model of TBI applies a pressure pulse directly on the intact dural surface by attaching a water-filled tube connected to a weight-driven piston to the skull (Fig. 1b) [7]. FPI models can deliver

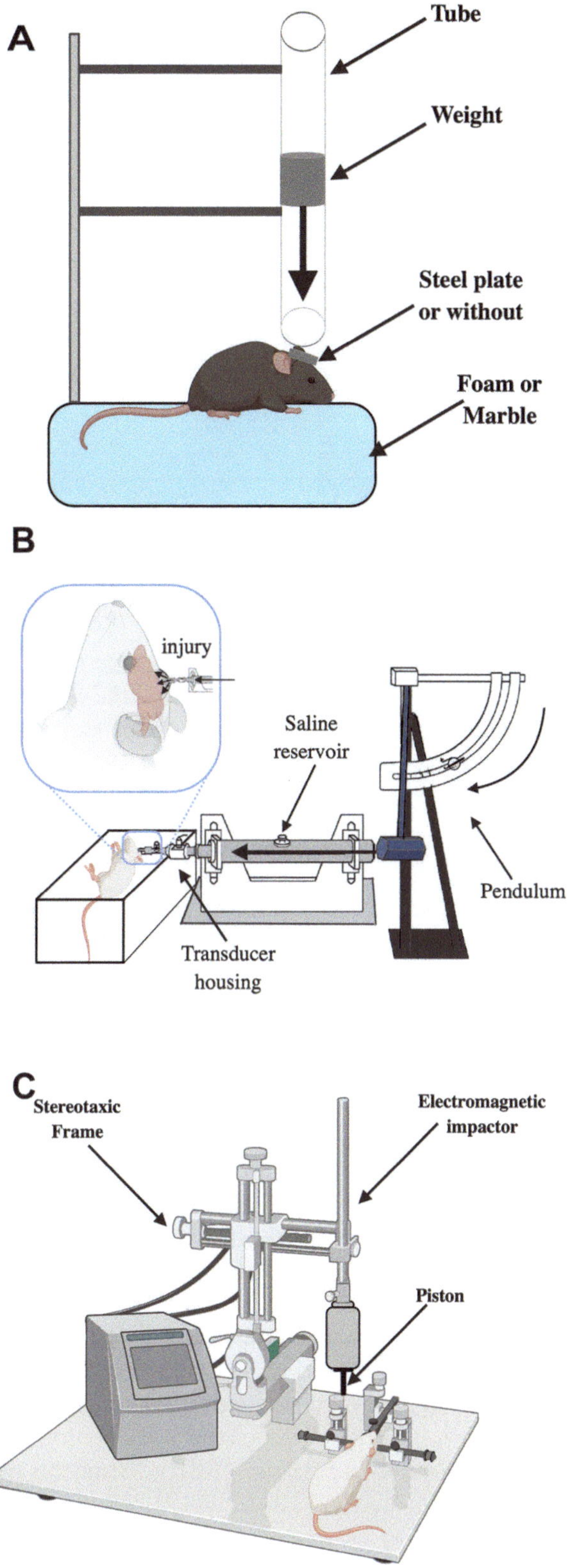

Fig. 1 Main models of TBI. (**a**) Weight-drop model. (**b**) Fluid percussion injury model. (**c**), Controlled cortical impact model

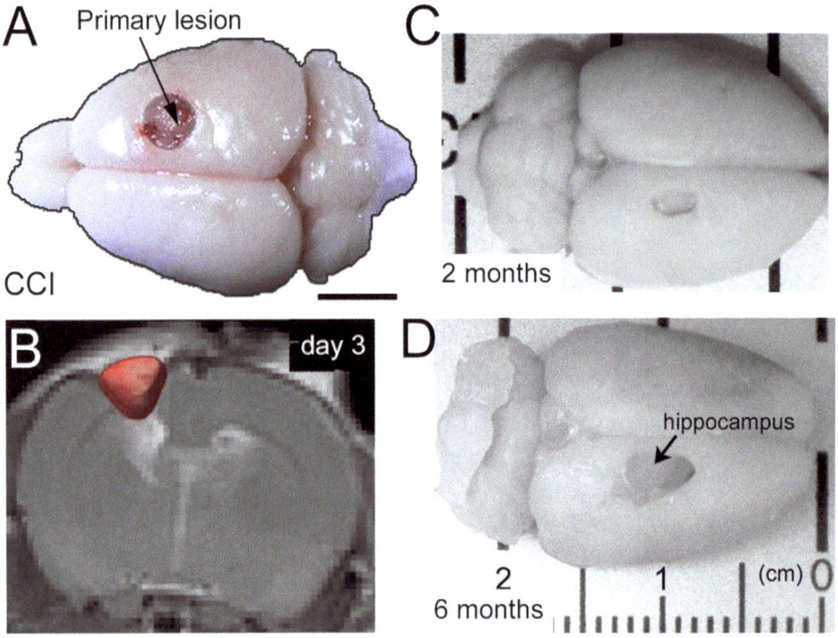

Fig. 2 (**a**) Picture representing an example of contusion with CCI model. In the center, the primary lesion with some bleeding and cell death. (**b**), T2WI at day 3 post-juvenile TBI (jTBI) with, in red, the 3D reconstructed lesion that gives a cavity at later time points postinjury. (**c**) and (**d**), Representative jTBI brains photographed postperfusion show the lesion as an open cavity spanning ~3 mm in length at 2 months and then ~6 mm to reveal the hippocampal formation (black arrow) beneath the missing cortex. (Modified from *Kamper* et al., *2013* [15])

various severities depending on the displacement of the pendulum (Fig. 1b) and where the pressure pulse is delivered on the brain: (1) laterally with a location of the cranial window over the parietal cortex (lateral FPI); or (2) onto the brain midline (central FPI). Each brain location provides a different subtype of TBI. Lateral FPI produces a mixed injury with cortical contusions combined with remote changes in ipsilateral hemisphere and white matter, whereas central FPI provokes diffuse injury with a wide variety of axonal injury, hippocampal damage, and brain stem lesion [7].

CCI models are widely used in the mouse and rat to induce focal cortical contusions with a range of severities at different developmental ages [10]. Below, we describe in detail the CCI method. CCI requires a craniotomy performed without damaging the dura so that a pneumatic or an electromagnetic piston can impact the exposed cortex with defined velocity, depth, and dwell time (Fig. 1c, detailed below). The impact to the cortex results in a cortical contusion associated with hemorrhage (if severe enough) and blood–brain barrier (BBB) disruption with almost immediate edema formation [10] (detailed below). After impact, the craniotomy has to be resealed in order to allow the development of intracranial hypertension [11]. The border of the hemorrhagic core undergoes necrosis at the site of the impact; the perilesional brain tissue exhibits a significant decrease of cerebral blood flow contributing to secondary injury cascades [12, 13]. They consist in a combination of events including vascular, cellular, and molecular processes, such as BBB disruption, edema formation, apoptosis, inflammation, and excitotoxicity during the initial 24–72 h after TBI. These secondary injuries progress from the core to adjacent primarily uninjured brain tissues [14, 15]. Over time, the contusion evolves into a cavity, which makes CCI a very good model to study pathophysiology as well as to test new therapeutic approaches (Fig. 2).

2 Materials

2.1 Anesthesia/Analgesia

Use a protocol in accordance with your local Institutional Animal Care and Use Committee (IACUC) to ensure the well-being of the animals.

1. Isoflurane (5% induction, 1–2% maintenance) (IsoVet, Piramal Healthcare, UK) is used for anesthesia. It is delivered through an isoflurane vaporizer and chamber for induction and then through a mask during the surgery (Fig. 3a–c). In our setup, the isoflurane vaporizer is connected with an 80 nitrogen/20 oxygen tank (Linde).
2. Buprenorphine (0.05 mg/kg s.c.) is injected subcutaneously (s.c.) in the mouse's back for analgesia.
3. Xylocaine (0.5%, 0.05 ml max) is applied for local anesthesia at the site of skin incision performed before the craniotomy.

2.2 Surgical Reagents

1. Saline 0.9% (NaCl).
2. Iodine soap.
3. Iodine 10% solution.
4. Surgical glue (cyanoacrylate).

2.3 Surgical Equipment (Fig. 3)

To maintain aseptic conditions, the entire surgical procedure is performed with surgical mask and sterile gloves. The tools are sterilized with cold sterilant (Actril), rinsed with deionized water and dried with 90% alcohol.

1. Scalpel blade, scalpel handle.
2. Fine forceps, tissue forceps.
3. Surgical microdrill (Nouvag, Switzerland) with corresponding drill bit (0.5–0.6 mm) is used to perform the cranial window (*see* Part 3.3).
4. Needle holder for suturing.

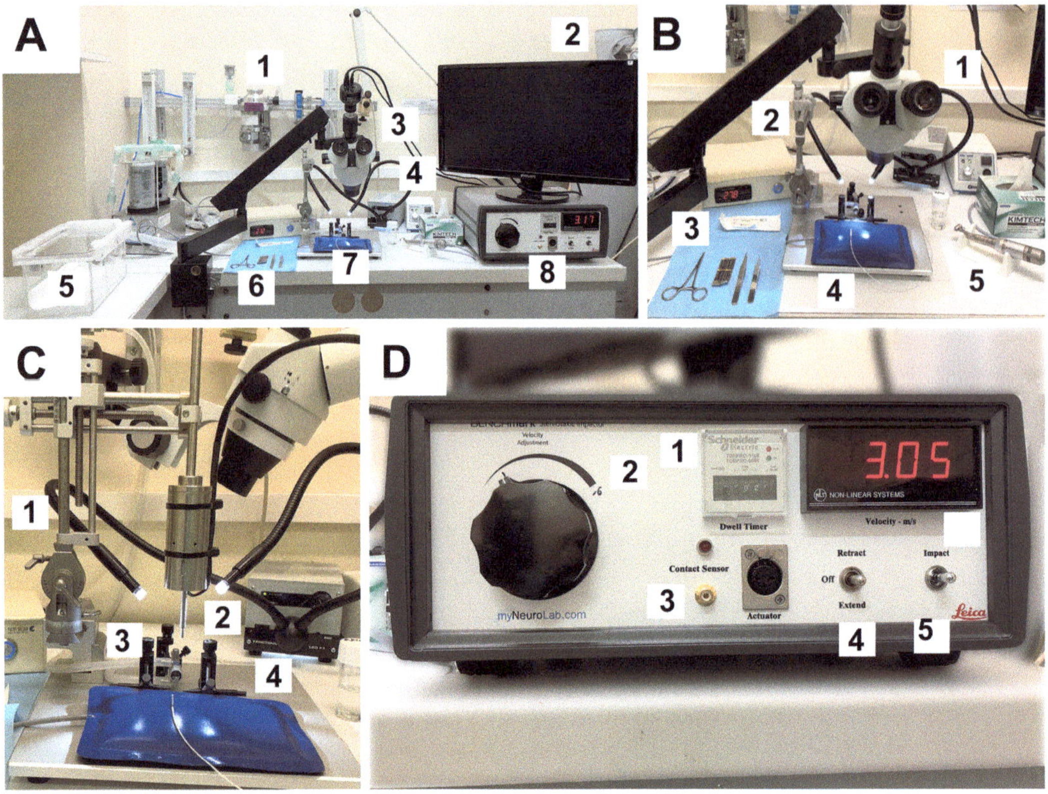

Fig. 3 Pictures of the CCI setup. (**a**) General overview of the surgery post. (**a1**) Isoflurane vaporizer connected to the gas tank (**a2**), the induction chamber (**a5**), and the mask (*see* **c3**). (**a3**) Binocular. (**a4**) Cold light source. (**a6**) Surgical tools easily reachable. (**a7**) Auto thermo-regulated heat pad. (**a8**) Impactor device. (**b**) Closer view of the surgical setup. (**b1**) Binocular. (**b2**) Stereotaxic arm. (**b3**) Surgical tools. (**b4**) Auto thermo-regulated heat pad. (**b5**) Drill. (**c**) Setup with impactor arm. (**c1**) Stereotaxic arm. (**c2**) Impactor tip. (**c3**) Anesthetic mask. (**c4**) Auto thermo-regulated heat pad. (**d**) Zoom on the impactor device. (**b1**) Dwell timer. (**b2**) Switch to adjust the impact velocity. (**b3**) Contact sensor light. (**b4**) Switch for the impactor tip position. (**b5**) Switch to make the impact

5. Suture: nonabsorbable 5-0 suture.
6. Cotton swabs, kimwipes, gauze sponges, or surgical sponges.
7. Cold light source to have enough light for conducting the surgical part.

2.4 Electromagnetic Impactor

The impactor used in this protocol is the electromagnetic impactor from Leica (Fig. 3a, d).

1. Impactor controller is set with chosen speed (6 m/s) on "OFF" position (Fig. 3d).
2. Impactor piston is fixed to the stereotaxic arm, with chosen tip, flat with 3 mm of diameter for this model. To increase the severity, the diameter of the tip can be wider with 5 mm or smaller with 1 mm.

2.5 Animal Maintenance Equipment

1. Heating pad at 37 °C, with rectal probe to maintain the temperature of the mouse during the surgery (Fig. 3).
2. Regular heating pads for the recovery period.
3. Eye protection using Vaseline.
4. Empty cage for recovery (half positioned on a heating pad).

For all animal use, it is required to have an approval of the experimentation protocol by your local and national ethical animal care and use committee (IACUC). For the work presented in this chapter, the CCI protocol has been approved under the Apafis number 19167 (University of Bordeaux).

3 Methods

A step-by-step protocol is detailed, and alternative approaches or modifications to the protocol are discussed in Subheading 4 and indicated by the corresponding note number.

3.1 Preparation Tools

Once all materials are collected, prepare your workstation in a comfortable way to have everything handy in order to perform the surgical procedure efficiently (Fig. 3a, b). Secure the impactor arm on the stereotaxic arm and set an angle of 20 degrees to have the impactor tip perpendicular to the cortical surface. This may vary between mice and rats, and also developmental age. Switch impactor "on" and set the speed at 6 m/s. The velocity of the tip can be adjusted depending on the injury severity desired [10]. Remove the stereotaxic arm holding the impactor tip from the stereotaxic frame so that you have easy access for prepping the craniotomy and store it close to you.

3.2 Preoperative Cares

In order to have a good aseptic condition, it is better to perform preoperative preparation in a different location than the surgery room.

1. After picking the animals from the animal facility, identify them with subcutaneous RFID chip for permanent identification or tail tattoo for long-term identification (over weeks) or permanent pen (couple of days). Weigh your animal and complete the printed monitoring datasheet (Fig. 4 for example) as you progress through the protocol.
2. Anesthetize the animal in a chamber prefilled with 5% isoflurane mixed in air.
3. Once animal is deeply anesthetized (absence of reflexes when pitching the hindlimb), place it on a mask at 1–2% isoflurane.
4. Inject buprenorphine subcutaneously (s.c. 0.05 mg/kg) (*see* **Note 1a**).

Monitoring datasheet for CCI animals

Surgeon	
Experiment	
Animal number	
Protocol	
Day post-surgery	
Date	

Weight	
% change	
Temperature	

Weight loss	**Score**	
< 5%	0	If score >1: check teeth, increase weighting frequency, make sweet and wet food available in cage
5 - 15%	1	
15 - 20%	2	
> 20%	3	

Apparency	
Normal	0
Lack of grooming, nasal or lacrimal runoff	1
Piloerection / Hunched back	3
Same + mid-closed eyes / spontaneous vocalisation	4

Behavior	
Normal	0
Less mobile and isolated but still attentive	1
Agitated / or highly immobile and not attentive	3

Provoked behaviour	
Normal	0
Slight changes: minor depression or exaggerated response	1
Violent reaction / Vocalisations / Weak reactions	3

Fig. 4 Example of monitoring datasheet to fill for postoperative follow-up

Clinical signs		
Normal	0	If temperature decreased of 1°C or more: place the cage on a warming pad. If wound: disinfection necessary
Moderate changes (wound, diarrhea, temperature +/- 1°C)	1	
Important changes (important bleedings, temperature +/- 2°C)	3	

Hydration		
Normal	0	If score 4: injection of warm lactate Ringer (1ml s.c., 2/day minimum)
Persistent skin fold	4	

TOTAL	/20

Treatment	
Rehydration: warm lactate Ringer 1ml (s.c. / i.p.)	
Buprenorphine: 0.05 mg/kg s.c.	
Attractive and wet sweet food (gello)	
Warming pad	

Depending on global score	
Normal	**0-5**
Careful observation: check clinical signs (temperature, teeth checking...) and pain signs, consider use of analgesic and increase follow-up frequency	**5-10**
Pain proven: buprenorphine complusory. Stopping the experimentation decided from case to case. Consider sacrifice if no improvement within 3 days.	**10-15**
Severe pain. Same than previously + stop the experimentation and sacrifice if the situation remains more than 3 days	**15-20**

Fig. 4 (continued)

5. Shave the head from behind the ears to the eye level and make sure not to damage ears and whiskers.
6. Clean the head thoroughly for the asepsis. We use the following protocol: three washes with iodine scrub followed by three applications of iodine solution.

7. Inject local anesthetic at the site of incision subcutaneously (xylocaine 0.5%, 0.05 ml max). Release the drug while withdrawing the needle, making a swollen line under the skin. Wait for a few minutes before making the incision.
8. Check the absence of reflexes before starting, by pinching the interdigital skin. Place the animal in the stereotaxic frame. There are different possibilities to place the mouse in the stereotaxic frame.

3.3 Craniotomy

1. Incise the skin on the top of the skull by making an incision long enough to be able to retract the skin off the impacted side (right side in our case). Remove periostea membrane with cotton swabs.
2. Dry the bone. Small blood leaks can be stopped by gently scratching the bone with the scalpel blade.
3. Draw the three sides of the window by scratching the bone with fine forceps (Fig. 5). The fourth side is the medial suture from which the bone flap is turned to leave an access to the cortical surface (Fig. 5). It is important to note that the prepared window must be wider than the impactor tip. Here, a 4-mm long window is made for a 3-mm diameter tip. To draw a reproducible window, a template can be used with the desired size such as a glass window. However, the window should not be performed above the assembly of the sagittal and transverse sinuses, to limit the risk of major bleeding when performing the craniotomy (Fig. 5).
4. We use an angled dental drill that can be held like a pencil with a 0.6-mm drill bit for a better precision of the movement (*see* **Note 2a**). Begin drilling the window along the marks (Fig. 5) with gentle movements over the different window sides. Do

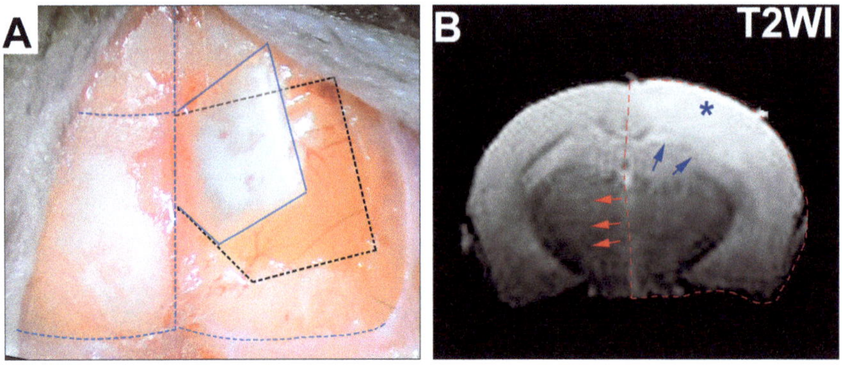

Fig. 5 (**a**) Representative picture of the cranial window before the injury with the bone flap. (**b**) T2WI MRI at 3 days post injury showing increase of the water content in the lesion (blue arrows and blue asterisk). The red dotted line is around the ipsilateral hemisphere, and the red arrows illustrate the mid-line shift indicating brain swelling

not push down; just scratch the surface of the bone without pressure on the drill. Pay attention to get the smallest width of window line.

5. Stop frequently the drilling to pour room temperature (RT) saline in order to prevent heating from the drilling, which may cause injury to the underlying cortical brain tissue. Wait a few minutes, dry the bone with swab and continue drilling. Do not drill completely the bone to avoid damaging the dura mater.
6. Gently push with forceps on the window edges of the bone in order to assess the progress of the drilling. The bone is thin enough when you feel that the bone flap moves down easily.
7. Once the three window edges are properly thinned, find a spot to grip the bone flap with thin forceps. Then, gently pull up the bone flap to determine if there is no remaining attachment on the prepared window. Do not force as you may injure the cortical surface. If the bone flap does not easily come up, go back to drilling. Make sure the area stays well hydrated with saline.

3.4 Injury

1. Once the window edges are properly drilled and the bone flap is set aside, set up the stereotaxic arm holding the impactor tip and secure it above the animal head.
2. Adjust the position of the tip at 20-degree angle in the window center at the surface of the bone. Then, move up the arm on the stereotaxic frame in order to extend the impactor (Fig. 3).
3. Gently flip the bone flap opened along the medial suture (*see* **Note 2b**). Check that there is no damage on the dura and no bleeding (*see* **Note 3**).
4. Set the impactor tip in the extended position (*see* **Note 4a** for impactor handling and Fig. 3d).
5. Lower the impactor tip using the stereotaxic descender until the impactor tip touches the dura surface. A sound from the impactor device indicates contact between dura and impactor tip (*see* **Note 4a**).
6. Then, set the impactor tip in the retracted position, which is approximately 10 mm.
7. Lower the stereotaxic arm with the impactor tip (Fig. 3d) at the desired depth. In this protocol designed for moderate CCI, a 1-mm depth is used. However, the depth can be changed to obtain various CCI severities [10].
8. Deliver the impact by switching the "Impact" knob on the impactor box (Fig. 3d). Pay attention that the impact is delivered properly in the center of the window without touching the bone flap during the impactor-tip course.

9. Immediately after the impact is completed, set the bone-flap back to its original place and glue it with surgical glue. Remove excess of glue once it is dried (*see* **Note 5** for variations of the model).
10. After the impact, the impactor tip returns to the retracted position. Put it back to the central "OFF" position to avoid overheating of the impactor device box (*see* **Note 4b**).
11. Close the skin with nonabsorbable suture. Let the animal to recover from anesthesia in a recovery cage placed on heating pad. Then, perform postoperative cares (*see* **Note 1d**).

3.5 Postsurgical Care

1. Weigh the animal and continue to fill the surgery datasheet.
2. Rehydrate the animal with s.c. injection of warm saline (0.1 ml/10 g).
3. Follow the analgesia protocol approved by your national and local animal ethics committee. In our approved protocol, we provide 2 days of 0.05 mg/kg buprenorphine s.c., but only if the animal is showing signs of pain.
4. Assess the injury and compare with the control group.

3.6 Control Group for CCI Model

Two different control groups are possible:

1. Sham-operated group corresponds to performing all steps for the craniotomy from Subheading 3.1–3.3. However, this group does not receive the impact, and therefore, steps in part 3.4 are not applied. Finally, steps in part 3.5 are followed until full recovery of the animal.
2. Naive group: The animal is placed under anesthesia for the same amount of time as that of animals from the CCI group, and it is then placed in recovery cage to wake up from anesthesia. No postsurgical care is given.

   The choice of the control group is discussed in **Note 6**. Various outcomes can be assessed to determine the extent and the severity of the injury after CCI protocol compared to sham-operated or naive groups.

3.7 CCI Outcome Measures

1. Diffusion MRI is a noninvasive neuroimaging that enables measurement of edema formation using a T2-weighted imaging (T2WI) sequence as well as lesion volumes over time in the same animal. It gives the opportunity to do gross morphological analysis with the measurement of midline shift due to brain swelling [10]. From T2 maps, T2 relaxation values can be recorded and compared between control and CCI groups. An increase in T2 values suggests water accumulation due to edema formation (Fig. 5b).

2. Histology can be carried out to measure tissue changes such as modifications of neuronal cell density using NeuN staining and blood–brain barrier leakage using IgG extravasation [16]. It can also help to evaluate inflammatory components with glial fribrillary acidic protein (GFAP) labeling for reactive astrocytes [16] and Ionize calcium-binding adaptor protein-1 (Iba1) staining for microglia [17].

4 Notes

1. Animal Care.

   In order to limit postoperative complications, specific care is needed during the surgical procedure to maintain animal's physiological parameters (temperature, ventilation, hydration, decreasing pain using analgesic, . . .).

   (a) *Anesthesia*: There is a synergic effect between buprenorphine and isoflurane that can lead to respiratory distress if doses are not monitored carefully. The isoflurane percentage should be maintained as low as possible; in addition, we do not use more than 0.05 mg/kg buprenorphine to sustain sufficient analgesia while keeping low the risk of respiratory distress. During the surgery, various vital signs of your animal should be monitored to help adjusting the anesthesia parameters, such as reflexes to pinch, respiration rate, and skin color.

   (b) *Hypothermia*: The animal body temperature drops as soon as the animal is anesthetized. Therefore, it is very important to provide a heating source using a heating pad and make sure to record the animal temperature during the surgery. In our protocol, we use standard heating pad for the pre- and postoperative care. However, a self-regulated heat pad with a rectal probe (Fig. 3) is used during the surgery to maintain the normal body temperature.

   (c) *Dehydration*: Hydration of the animal is checked by pinching the skin over the shoulder blades after 30 min of surgery. In a well-hydrated animal, the skin will return quickly to its original position. If it does not return to a normal position, or if major bleeding occurred during the surgery procedure, the animal is likely to become dehydrated. Therefore, warm saline should be injected s.c. with 0.1 ml for 10 g of body weight. At the end of the surgery, injection of warm saline should be performed in order to facilitate the recovery of the animals.

   (d) *Postsurgery Monitoring*: Animals are monitored for minimum 3 days after surgery. We use a monitoring datasheet

(Fig. 4) for each animal, which includes weight, hydration (skin pinch test), behavior (pain assessment), and observation of the wound. In the moderate CCI model presented in this chapter, most animals do not need particular care as long as the physiological parameters are kept normal during surgery. The animals usually exhibit normal behavior 24 h after the end of the surgery.

2. Craniotomy and Bleeding.
   (a) Bone bleeding can occur while drilling the edges of the cranial window. This bleeding should be stopped prior to continued drilling. We found that faster drilling decreases bleeding. Therefore, the drilling speed should be between 10,000 and 20,000 rpm to reduce the chance of bone bleeding.
   (b) Attention needs to be paid during drilling to avoid damage of microvessels and cortical surface. Hemorrhage due to the damage of pial blood vessels during the drilling process increases the severity of the trauma. If by accident there is a micro hemorrhage, try to stop it with cotton swab and pour room temperature saline. You should remove the animal from the CCI cohort and report on the surgery sheet any animal that experienced major bleeding during the craniotomy.
   (c) Flipping the cranial window along the midline can also induce bleeding. In order to limit any bleeding when the bone flap is moved, pay a special attention to drill up to the skull suture, so that any movement of the bone flap will be along brain skull suture (Fig. 5).

3. Dura Mater Lesion During Craniotomy.

   We typically exclude animals with dura lesion during the cranial window preparation. The impact delivered by CCI apparatus (1 mm depth and 6 m/s) does not induce a systematic dura mater rupture, and it is considered as a moderate CCI. However, including animals with damaged dura is acceptable for a model of severe CCI, where the dura will be broken.

4. Impactor Issues.
   (a) *Manipulation*: Before starting the surgery and CCI protocol, be sure that you are familiarized with the use of the electromagnetic impactor device. Pay a special attention to set the impactor tip in "extend" position before adjusting it on the skull surface. It is important to note that no contact noise will appear when the tip is on any other position than "extend."
   (b) *Overheating*: The impactor electronic box is very sensitive to overheating. It is mandatory to switch back the

impactor tip on the position "OFF" after the impact delivery. Any other position should not be held for more than 7 min; otherwise, there is a risk of overheating, leading to a nonaccurate velocity.

(c) *Device Accuracy*: To ensure accuracy along the CCI procedure, it is important to check the proper screwing of the tip and the impactor arm often as they have tendency to loosen upon use.

5. Variations in the CCI Model.

   (a) The Craniotomy Left Open:

   - The square bone flap is left attached to the medial suture in order to close quickly the craniotomy. This maintains physical constrains of the skull and reproduces the increase in intracranial pressure (ICP) when the brain swells, as it was originally described [18].
   - One existing variation is to leave the craniotomy opened. This variation has been described in the literature, showing no tissue growth outside the skull [19]. This variation gives the opportunity to score the severity of the injury by quantifying swelling and bleeding (Fig. 6). However, ICP is reduced or blunted, and then, fewer secondary injuries may occur.

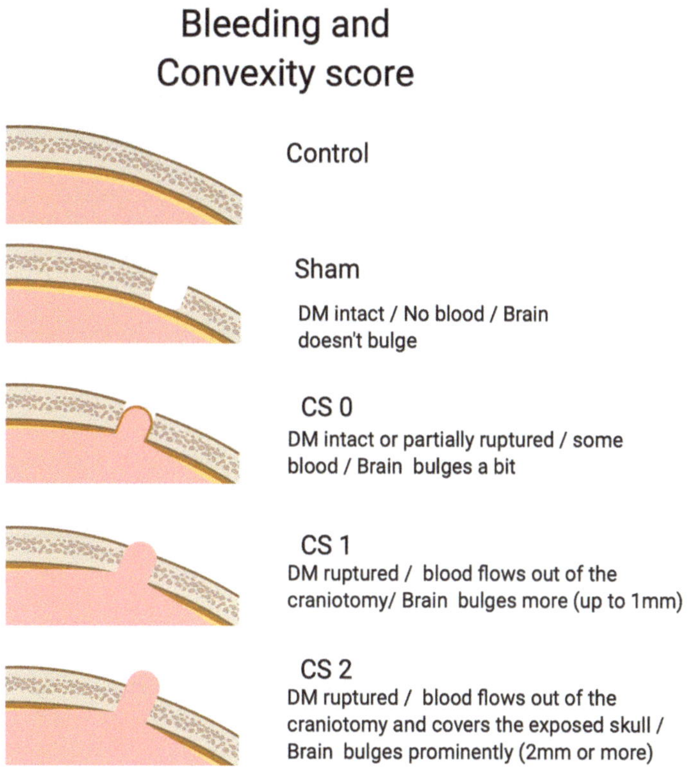

Fig. 6 Scoring the severity of the injury by quantifying swelling and bleeding

(b) *Round Cranial Window*:
- This cranial window can also be prepared with a trephine mounted on the drill apparatus. It is difficult to return and secure quickly a bone flap that has been completely removed. Therefore, this approach is better suited for surgeries where the cranial window is not closed or with the use of a bone substitute.

(c) *Severity Score*:
- The following parameters 6 m/s speed and 1-mm depth give a moderate injury on mouse cortex. The electromagnetic impactor device delivers impacts with a speed between 1 and 6 m/s. It is also possible to vary the depth of impact as well. By changing speed and depth, it allows severity variations of CCI. This protocol has been optimized for young adult C57BL6 mice (12-week-old). This CCI protocol can be used on mouse and rat with minor modifications as needed. It can also be performed in juvenile rats, as in our group we used CCI in juvenile rats with various severities [10, 20].

6. Appropriate Controls.
   (a) We consider the appropriate control being naive animals since we are interested in the difference between injured brain and normal brain. We believe that sham-operated animals, even if the surgery was perfectly performed, replicative of normal brain, and we consider the craniotomy surgery being part of the TBI. Naive animals must experience the same length of anesthesia as that of CCI animals.
   (b) Sham-operated animals can be used to detect inflammation and damage caused only by the disruption of the skull, to see whether it led to infection, and exclude it from the mechanical effect of the TBI.

Acknowledgments

The authors declare to have no conflict of interests. Fig. 1 and 6 have been partly made using www.biorender.com. This study was supported by grants eraNET neuron CNSAflame (JB), TRAINS (JB), and Direction Générale des Armées (TC, JB).

References

1. Faul M, Xu L, Wald MM et al. (2010) Traumatic brain injury in the United States: emergency department visits, hospitalizations, and deaths, 2002–2006. . National Center for Injury Prevention and Control Atlanta, GA: CDC
2. Selassie AW, Zaloshnja E, Langlois JA et al (2008) Incidence of long-term disability following traumatic brain injury hospitalization,

United States, 2003. J Head Trauma Rehabil 23(2):123–131. https://doi.org/10.1097/01.HTR.0000314531.30401.39

3. Zaloshnja E, Miller T, Langlois JA et al (2008) Prevalence of long-term disability from traumatic brain injury in the civilian population of the United States, 2005. J Head Trauma Rehabil 23(6):394–400. https://doi.org/10.1097/01.HTR.0000341435.52004.ac
4. Johnson VE, Stewart W, Smith DH (2010) Traumatic brain injury and amyloid-beta pathology: a link to Alzheimer's disease? Nat Rev Neurosci 11(5):361–370. https://doi.org/10.1038/nrn2808
5. Coronado VG, Xu L, Basavaraju SV et al (2011) Surveillance for traumatic brain injury-related deaths--United States, 1997-2007. MMWR Surveill Summ 60(5):1–32
6. Pop V, Badaut J (2011) A neurovascular perspective for long-term changes after brain trauma. Transl Stroke Res 2(4):533–545. https://doi.org/10.1007/s12975-011-0126-9
7. Marklund N, Hillered L (2011) Animal modelling of traumatic brain injury in preclinical drug development: where do we go from here? Br J Pharmacol 164(4):1207–1229. https://doi.org/10.1111/j.1476-5381.2010.01163.x
8. Foda MA, Marmarou A (1994) A new model of diffuse brain injury in rats. Part II: morphological characterization. J Neurosurg 80(2):301–313. https://doi.org/10.3171/jns.1994.80.2.0301
9. Marmarou A, Foda MA, van den Brink W et al (1994) A new model of diffuse brain injury in rats. Part I: pathophysiology and biomechanics. J Neurosurg 80(2):291–300. https://doi.org/10.3171/jns.1994.80.2.0291
10. Badaut J, Adami A, Huang L et al (2019) Noninvasive magnetic resonance imaging stratifies injury severity in a rodent model of male juvenile traumatic brain injury. J Neurosci Res 98(1):129–140. https://doi.org/10.1002/jnr.24415
11. Zweckberger K, Plesnila N (2009) Anatibant, a selective non-peptide bradykinin B2 receptor antagonist, reduces intracranial hypertension and histopathological damage after experimental traumatic brain injury. Neurosci Lett 454(2):115–117. https://doi.org/10.1016/j.neulet.2009.02.014
12. Hayward NM, Immonen R, Tuunanen PI et al (2010) Association of chronic vascular changes with functional outcome after traumatic brain injury in rats. J Neurotrauma 27(12):2203–2219. https://doi.org/10.1089/neu.2010.1448
13. Engel DC, Mies G, Terpolilli NA et al (2008) Changes of cerebral blood flow during the secondary expansion of a cortical contusion assessed by 14C-iodoantipyrine autoradiography in mice using a non-invasive protocol. J Neurotrauma 25(7):739–753. https://doi.org/10.1089/neu.2007.0480
14. Michinaga S, Koyama Y (2015) Pathogenesis of brain edema and investigation into anti-edema drugs. Int J Mol Sci 16(5):9949–9975. https://doi.org/10.3390/ijms16059949
15. Kamper JE, Pop V, Fukuda AM et al (2013) Juvenile traumatic brain injury evolves into a chronic brain disorder: behavioral and histological changes over 6 months. Exp Neurol 250:8–19. https://doi.org/10.1016/j.expneurol.2013.09.016
16. Fukuda AM, Pop V, Spagnoli D et al (2012) Delayed increase of astrocytic aquaporin 4 after juvenile traumatic brain injury: possible role in edema resolution? Neuroscience 222:366–378. https://doi.org/10.1016/j.neuroscience.2012.06.033
17. Fukuda AM, Adami A, Pop V et al (2013) Posttraumatic reduction of edema with aquaporin-4 RNA interference improves acute and chronic functional recovery. J Cereb Blood Flow Metab 33(10):1621–1632. https://doi.org/10.1038/jcbfm.2013.118
18. Hamm RJ, Dixon CE, Gbadebo DM et al (1992) Cognitive deficits following traumatic brain injury produced by controlled cortical impact. J Neurotrauma 9(1):11–20
19. Romine J, Gao X, Chen J (2014) Controlled cortical impact model for traumatic brain injury. J Vis Exp 90:e51781. https://doi.org/10.3791/51781
20. Ajao DO, Pop V, Kamper JE et al (2012) Traumatic brain injury in young rats leads to progressive behavioral deficits coincident with altered tissue properties in adulthood. J Neurotrauma 29(11):2060–2074. https://doi.org/10.1089/neu.2011.1883

Chapter 7

Evaluation of MicroRNA Therapeutic Potential Using the Mouse In Vivo and Human Ex Vivo Wound Models

Xi Li and Ning Xu Landén

Abstract

Wound healing is a fundamental physiological process to keep the integrity of the skin; failure of wound healing leads to chronic wounds, which are a common and severe medical problem. MicroRNAs (miRNAs) are gene regulators important for multiple biological functions in the skin, and they play essential roles in different phases of wound repair. Many miRNAs have been found dysregulated in human chronic wounds. Therefore, miRNAs may serve as potential therapeutic targets for wound treatment. In this chapter, we describe a step-by-step protocol about how to evaluate the therapeutic potential of a miRNA in mouse in vivo and human ex vivo wound models. The findings from these preclinical wound models will serve as a basis for further clinical trials.

Key words microRNA mimic, microRNA inhibitor, Chronic wounds, Mouse in vivo wound model, Human ex vivo wound model

1 Introduction

Skin is an essential biological barrier of the human body, and wound healing is the fundamental physiological process to keep the integrity of the skin. However, for the patients with venous or arterial insufficiency, patients with diabetes, and patients with continuous pressure of the bodyweight to the skin (e.g., spinal cord injury patients and the bedridden elderly population), their wounds are often difficult to heal and form chronic nonhealing wounds such as venous/arterial ulcer, diabetic foot ulcer, and pressure ulcer. Chronic wound is a growing socioeconomic and health concern, which longs for a deeper understanding of its pathophysiology to discover more effective wound treatments.

MicroRNAs (miRNAs) are approximately 22 nucleotides long noncoding RNAs, which could repress translation or degrade their target mRNAs [1]. MiRNAs play essential regulatory roles in almost all known biological processes, and their dysregulation has been shown to contribute to multiple pathological conditions, such

Hiranmoy Das (ed.), *Wound Regeneration: Methods and Protocols*, Methods in Molecular Biology, vol. 2193, https://doi.org/10.1007/978-1-0716-0845-6_7,

as cancer, cardiovascular diseases, infection, and inflammatory diseases. Emerging clinical trials have revealed that the administration of miRNA-specific mimics or inhibitors to modulate miRNA expression shows promising therapeutic effects on multiple diseases, including virus infection, cancer, and diabetes [2, 3].

Can we treat wounds by targeting miRNAs? For this, we need to understand the expression and functions of miRNAs in wound healing. A miRNA expression profile of human wounds has been reported [4]. Among the miRNAs changed during healing process, many have been demonstrated to be important for wound repair, for example, miR-21 [5], miR-27b [6], miR-31 [7], miR-99 family [8], miR-132 [4, 9, 10], miR-146a [11], miR-155 [12], miR-210 [13], and miR-34 family [14]. Interestingly, miR-132 has been shown downregulated in human diabetic foot ulcer [9], whereas miR-21, miR-34a, and miR-34b were found upregulated in human venous ulcer [15, 14]. These miRNAs may serve as therapeutic targets to treat chronic wounds.

To test the therapeutic potential of miRNAs, both mice in vivo and human ex vivo wound models have been used [9, 15]. Even though there are differences between murine and human skin repair, the mouse model still shares the main features of human wound healing and is the most commonly used in vivo wound model [16]. Moreover, the clinical features of chronic wounds can be recapitulated in different mice models, such as a leptin-deficiency (db/db) mouse, which is a model of type 2 diabetes mellitus, and it exhibits delayed wound healing [17, 18]. To complement the data obtained in an animal model, the human ex vivo wound model allows us to determine whether interference with miRNA expression in a complex setting can improve the re-epithelialization of human wounds [19]. In this chapter, we describe how to locally overexpress or inhibit a specific miRNA in mouse in vivo and human ex vivo wound models to test its therapeutic potential.

2 Materials

2.1 Mouse In Vivo Wound Model

1. Twelve-week-old male wild-type C57BL/6N mice (WT mice) or diabetic BKS(D)-Leprdb/JOrlRj mice on the C57Bl/6J background (db/db mice).
2. Electric clippers.
3. Hair removal (depilatory) cream.
4. Small-animal anesthesia system.
5. Isoflurane.
6. Buprenorphine: Temgesic® 1 mL/tube diluted in 30-mL sterile phosphate-buffered saline (PBS) (pH 7.4) under biosafety hood. Keep it cold (+4 °C) before use. Inject intraperitoneally 100 μL/10 g body weight.

7. Surgical equipment: 4-mm biopsy punch, tissue forceps, anatomical forceps, and curved and straight scissors.
8. MaxSuppressor™ In Vivo RNA-LANCEr II and Lipid Extruder (Bioo Scientific).
9. mirVana miRNA mimics specific to the studied miRNA and miRNA-negative control (ThermoFisher Scientific) or miRCURY LNA miRNA inhibitors and negative control (QIAGEN) (*see* **Note 1**).
10. Sterile RNAse-free water.
11. 1× and 10× PBS (pH 7.4).
12. Digital camera.

2.2 Human Ex Vivo Wound Model

1. Dulbecco's modified Eagle's medium (DMEM) with 10% fetal bovine serum (FBS) and 1% PEST (penicillin 100 units/L and streptomycin 100 μg/mL; ThermoFisher Scientific).
2. Surgical instruments: sterile skin biopsy punch (2 mm and 6 mm), tissue forceps, anatomical forceps, and curved and straight scissors.
3. 70% ethanol.
4. Sterile 1× PBS.
5. MaxSuppressor™ In Vivo RNA-LANCEr II and Lipid Extruder (Bio Scientific).
6. 30% pluronic F-127 gel (Sigma-Aldrich).
7. mirVana miRNA mimics specific to the studied miRNA and miRNA-negative control (ThermoFisher Scientific) or miRCURY LNA miRNA inhibitors and negative control (QIAGEN) (*see* **Note 1**).

3 Methods

3.1 Mouse In Vivo Wound Model

1. Before surgery, mice are caged individually for 1 week and handled daily (*see* **Note 2**).
2. One day before wounding, general anesthesia of the mice is performed with 3% isoflurane for at least 5 min using a small-animal anesthesia system. Then, the hairs on the back of mice are shaved with an electronic clipper followed by an application of depilatory cream and then washed with 1× PBS. The weight of each mouse is measured.
3. On the day of surgery, general anesthesia is performed as described above. Wipe the skin with 1× PBS. Two full-thickness wounds extending through the panniculus carnosus are made on the dorsum on each side of the midline using a 4-mm biopsy punch.

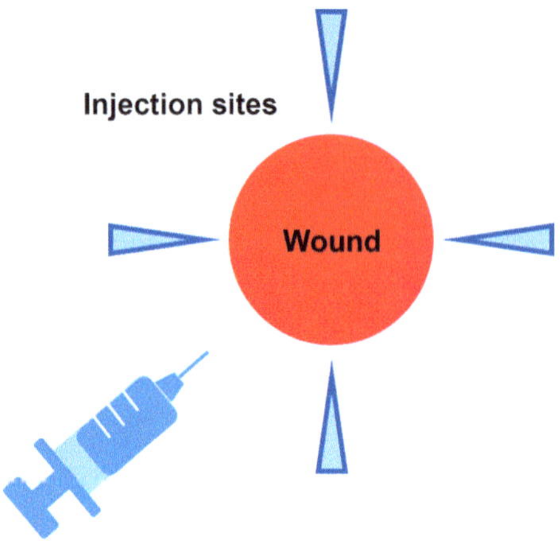

Fig. 1 Intradermal injection of miRNA mimics or inhibitors into four sites around the wound edge

4. The wound can be treated with miRNA mimics or inhibitors (*see* **Note 1**) immediately after the injury or at the indicated time points during wound healing.

   A 100-μL mixture of miRNA mimics and MaxSuppressor™ In Vivo RNA-LANCEr II is injected intradermally into four sites (approximately 25 μL per location) at the edge of each wound (Fig. 1) (*see* **Notes 3** and **4**). This mixture is prepared as described below:

   Add 15 nmol of miRNA mimics or negative control mimic plus 10% overage (i.e., 16.5 nmol) into a vial of MaxSuppresor™ In Vivo RNA-LANCEr. If 0.5 nmol of miRNA mimics in a total volume of 100 μL will be injected into each wound, 300 μL of RNase-free 10× PBS and 2550 μL of RNAse-free water will be added into the same vial. This results in a mixture of 3000 μL, which is enough to treat about 27 wounds. Disperse the liposome emulsion in aqueous solution using either the Lipid Extruder, which sizes liposomes to 100 nm, or a water bath sonicator, which sizes liposomes to 15–100 nm. Sonication is performed by immersing the vial into the water bath sonicator for 5 min on a high-power setting. This mixture can be aliquoted and stocked in −80 °C freezer before use.

   For inhibition of miRNA expression, MaxSuppresor™ In Vivo RNA-LANCEr is not needed. One nanomole miRNA-specific inhibitors or negative control dissolved in 100 μL RNase free 1× PBS is injected intradermally into the edge of each wound (*see* **Note 3**).

5. Photograph each wound with a scale next to it to document its size.

6. After the surgical procedure, mice receive intraperitoneal buprenorphine for relief of pain and distress.

7. The mice are caged individually.
8. The wound areas are photographed with a scale every other day until euthanized (*see* **Note 5**). The wound size is measured using ImageJ 1.32 software (National Institutes of Health) and calculated as the percentage of its initial size.
9. Mice will be euthanized with carbon dioxide at the specified time points after injury, and the following biopsies are collected: wound edges as well as the skin near and far away from the wounds; and inner organs, for example, liver, lung, kidney, and spleen. These samples can be photographed to document their morphology, and then quickly frozen on dry ice.
10. The following analysis may be performed to evaluate the effects of miRNA-based treatments on wound healing:
    (a) Quantification of wound-healing speed macroscopically (*see* Subheading 3.1, **step 8**) and histologically by hematoxylin and eosin (H&E) staining of the wound tissues.
    (b) Detection of miRNA overexpression or inhibition, as well as its target gene expression by real-time qRT-PCR, in situ hybridization, and immunostaining of the treated wound tissues.
    (c) Depending on the functions of the studied miRNAs, one could check whether these biological processes are altered after miRNA treatment. For example, cell proliferation may be analyzed by immunostaining of proliferation marker MKI67 in wound biopsies; immune cell infiltration may be evaluated by immunostaining of specific immune cell markers.
    (d) The potential systemic adverse effects may be evaluated by observation of the macro morphology of inner organs; H&E staining of tissue sections of the inner organs; quantification of the studied miRNA level in the skin distal and proximal to the wounds, and in inner organs by real-time qRT-PCR.

3.2 Human Ex Vivo Wound Model

1. Human skin tissues are obtained from surgeries such as abdominal reduction operation, transported on ice to the lab, and processed on the same day. All skin donors provide informed consent for the study. The study must be approved by the local ethics committee and is conducted according to the Declaration of Helsinki Principles.
2. In a cell culture hood, the skin surface is wiped with tissue papers to remove blood, and then cleaned with 70% ethanol and sterile 1× PBS.

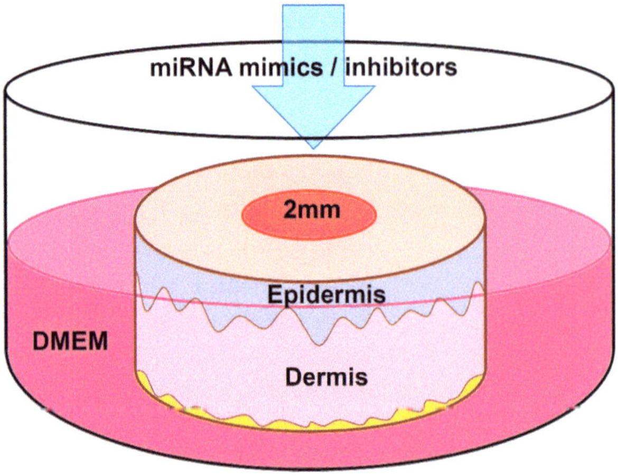

Fig. 2 The human ex vivo wound model is cultured in DMEM

3. A superficial wound is created using a 2-mm biopsy punch on the skin (*see* **Note 6**), and then, the wound is excised from the skin by using a 6-mm biopsy punch. Remove subcutaneous fat with a pair of scissors, and then transfer the tissue to a 12-well cell culture plate.
4. 800 μL DMEM with 10% FBS and antibiotics is added around the skin tissue, and let its epidermal surface be exposed in the air to create a liquid–air interface (Fig. 2).
5. 0.1 nmol mirVana miRNA mimic or negative control is packed in transfection reagent MaxSuppressor™ In Vivo RNA-LANCEr II (*see* **step 4** in Subheading 3.1), and then dissolved in 10 μL 30% pluronic F-127 gel. For inhibition of miRNA expression, MaxSuppresor™ In Vivo RNA-LANCEr is not needed. 0.5 nmol miRNA-specific inhibitors or negative control is dissolved in 10 μL 30% pluronic F-127 gel. Pipette 10-μL mixture of miRNA mimics/inhibitor with pluronic F-127 gel onto each wound (*see* **Notes 3** and 7).
6. Human ex vivo wounds usually re-epithelize completely in about 4–7 days. The culture medium can be changed once or twice during this process. Wound samples can be collected at multiple time points postinjury: put the wound tissues on a piece of tissue paper to remove culture medium, and then frozen them on dry ice. The tissues can be used for gene expression and histological analysis of the expression of the studied miRNA and its targets. Skin re-epithelialization can be evaluated by H&E staining, and the lengths of the newly formed epithelial tongues are measured (*see* **Note 8**).

4 Notes

1. mirVana miRNA mimics are used to overexpress the studied miRNAs, whereas miRCURY LNA miRNA inhibitors are used to suppress the function of the studied miRNAs.
2. The mice need to be individually caged after wounding since they tend to scratch and bite each other's wounds. And to reduce the psychological pressure after the surgery, the mice are individually caged 1 week before the operation to allow them to get used to the living environment.
3. Doses of miRNA mimics and inhibitors used in wound treatment and frequency of treatments may need to be optimized.
4. Intradermal injection will result in the formation of a small and restricted liquid bulge. If the liquid is injected subcutaneously, the bulge area will not have clear edges and will disappear quickly.
5. Scabs are removed before taking photographs to expose the wound edge when analyzing wound healing macroscopically. If the wound tissue is collected for histological analysis, the scab removing process will be omitted, as it will disturb the newly formed epithelial tongue.
6. The depth of the wound should be approximately 1 mm, which includes the full thickness of epidermis and the upper part of the dermis. One could mark the biopsy punch 1 mm above the blade, to standardize the wound depth (Fig. 3).

Fig. 3 Mark biopsy punch 1 mm above the blade with a red line

7. 30% pluronic F-127 gel forms a free-flowing solution at or below normal ambient temperature, and it is able to form a gel at body temperature. Once the reagents are mixed with pluronic F-127 gel, keep the vial on ice before pipetting the mixture onto the wound tissue.
8. To measure the size of the wound accurately, each wound tissue is cut in half and sectioned across the mid-line of the wound.

Acknowledgments

We would like to thank Dr. Dongqing Li for critical reading of the chapter. This work is supported by Swedish Research Council (Vetenskapsradet, Dnr. 2016-02051), Ragnar Söderbergs Foundation (M31/15), Welander and Finsens Foundation (Hudfonden), Ming Wai Lau Centre for Reparative Medicine, Karolinska Institutet, and the Chinese Scholarship Council (a Ph.D. scholarship to XL).

References

1. Jonas S, Izaurralde E (2015) Towards a molecular understanding of microRNA-mediated gene silencing. Nat Rev Genet 16 (7):421–433. https://doi.org/10.1038/nrg3965
2. Chakraborty C, Sharma AR, Sharma G et al (2017) Therapeutic miRNA and siRNA: moving from bench to clinic as next generation medicine. Mol Ther Nucleic Acids 8:132–143. https://doi.org/10.1016/j.omtn.2017.06.005
3. Rupaimoole R, Slack FJ (2017) MicroRNA therapeutics: towards a new era for the management of cancer and other diseases. Nat Rev Drug Discov 16(3):203–222. https://doi.org/10.1038/nrd.2016.246
4. Li D, Wang A, Liu X et al (2015) MicroRNA-132 enhances transition from inflammation to proliferation during wound healing. J Clin Invest 125(8):3008–3026. https://doi.org/10.1172/JCI79052
5. Das A, Ganesh K, Khanna S et al (2014) Engulfment of apoptotic cells by macrophages: a role of microRNA-21 in the resolution of wound inflammation. J Immunol 192 (3):1120–1129. https://doi.org/10.4049/jimmunol.1300613
6. Wang JM, Tao J, Chen DD et al (2014) MicroRNA miR-27b rescues bone marrow-derived angiogenic cell function and accelerates wound healing in type 2 diabetes mellitus. Arterioscler Thromb Vasc Biol 34(1):99–109. https://doi.org/10.1161/ATVBAHA.113.302104
7. Li D, Li XI, Wang A et al (2015) MicroRNA-31 promotes skin wound healing by enhancing keratinocyte proliferation and migration. J Invest Dermatol 135(6):1676–1685. https://doi.org/10.1038/jid.2015.48
8. Jin Y, Tymen SD, Chen D et al (2013) MicroRNA-99 family targets AKT/mTOR signaling pathway in dermal wound healing. PLoS One 8(5):e64434. https://doi.org/10.1371/journal.pone.0064434
9. Li X, Li D, Wang A et al (2017) MicroRNA-132 with therapeutic potential in chronic wounds. J Invest Dermatol 137 (12):2630–2638. https://doi.org/10.1016/j.jid.2017.08.003
10. Li X, Li D, Wikstrom JD et al (2017) MicroRNA-132 promotes fibroblast migration via regulating RAS p21 protein activator 1 in skin wound healing. Sci Rep 7(1):7797. https://doi.org/10.1038/s41598-017-07513-0
11. Roy S, Elgharably H, Sinha M et al (2014) Mixed-species biofilm compromises wound healing by disrupting epidermal barrier function. J Pathol 233(4):331–343. https://doi.org/10.1002/path.4360
12. van Solingen C, Araldi E, Chamorro-Jorganes A et al (2014) Improved repair of dermal wounds in mice lacking microRNA-155. J

Cell Mol Med 18(6):1104–1112. https://doi.org/10.1111/jcmm.12255

13. Biswas S, Roy S, Banerjee J et al (2010) Hypoxia inducible microRNA 210 attenuates keratinocyte proliferation and impairs closure in a murine model of ischemic wounds. Proc Natl Acad Sci U S A 107(15):6976–6981. https://doi.org/10.1073/pnas.1001653107
14. Wu J, Li X, Li D et al (2019) MicroRNA-34 family enhances wound inflammation by targeting LGR4. J Invest Dermatol 140 (2):465–476.e11. https://doi.org/10.1016/j.jid.2019.07.694
15. Pastar I, Khan AA, Stojadinovic O et al (2012) Induction of specific microRNAs inhibits cutaneous wound healing. J Biol Chem 287 (35):29324–29335. https://doi.org/10.1074/jbc.M112.382135
16. Zomer HD, Trentin AG (2018) Skin wound healing in humans and mice: challenges in translational research. J Dermatol Sci 90 (1):3–12. https://doi.org/10.1016/j.jdermsci.2017.12.009
17. Burke SJ, Batdorf HM, Burk DH et al (2017) db/db mice exhibit features of human type 2 diabetes that are not present in weight-matched C57BL/6J mice fed a Western diet. J Diabetes Res 2017:8503754. https://doi.org/10.1155/2017/8503754
18. Kobayashi K, Forte TM, Taniguchi S et al (2000) The db/db mouse, a model for diabetic dyslipidemia: molecular characterization and effects of Western diet feeding. Metabolism 49(1):22–31. https://doi.org/10.1016/s0026-0495(00)90588-2
19. Heilborn JD, Nilsson MF, Kratz G et al (2003) The cathelicidin anti-microbial peptide LL-37 is involved in re-epithelialization of human skin wounds and is lacking in chronic ulcer epithelium. J Invest Dermatol 120(3):379–389. https://doi.org/10.1046/j.1523-1747.2003.12069.x

Chapter 8

Cellular Migration Assay: An In Vitro Technique to Simulate the Wound Repair Mechanism

A K M Nawshad Hossian and George Mattheolabakis

Abstract

Wound regeneration is a complex process, which necessitates proper coordination among the inflammatory response, vascularization, matrix formation, and reformation of epithelial tissue. It is a unique process, where healing and regeneration take place simultaneously. Matrix formation is the first critical stage that starts the communication between the keratinocytes, fibroblasts, and integrins. This, in turn, stimulates the differentiation of monocytes into macrophages, to produce cytokines for fibroblasts. This phenomenon is the crucial part for the keratinocyte migration and epithelialization to fill the wound. To understand the complex procedure of wound regeneration, there is a need for easy, convenient, and low-cost approaches that will simulate the wound-repairing process. Scratch assay or cellular migration assay is one of the most convenient and affordable approaches, commonly used by the scientific community. In this chapter, we present the fundamental principles of the experimental procedures required for the Scratch assay.

Key words Scratch assay, Transfection, Wound-healing assay, ImageJ

1 Introduction

The basic principle behind wound-healing and scratch assay is to study the rate of closure of an artificially created gap by a monolayer of cells (filling-up the space). The artificial gap is created by scratching off cells from the monolayer. This procedure is a well-established in vitro approach to understand the basic concepts of cellular migration to wound recovery [1]. In vivo wound regeneration demonstrates a complex behavior, where the involvement of immune cells, like monocytes, dendritic cells, neutrophils, and lymphocytes, plays critical roles along with endothelial cells, keratinocytes, and fibroblasts. There is a series of genetic changes taking place among different cell types to potentiate the repair mechanism [2, 3]. This process also involves the activation of several cellular pathways, including HB-EGF, TGFβ, and MAPK [4]. Though scratch assay does not include the involvement of immune responsive cells and other complex cellular behaviors present in the

Hiranmoy Das (ed.), *Wound Regeneration: Methods and Protocols*, Methods in Molecular Biology, vol. 2193,
https://doi.org/10.1007/978-1-0716-0845-6_8,

complicated natural multicellular environment, the relatively low cost and easy and fast experimental procedure have made this assay attractive to researchers.

It is currently well understood that microRNAs (miRs) are genetic regulatory molecules, which are involved in regulating gene expressions and pathways of different disease conditions [5]. miRs are involved in a plethora of natural mechanisms(e.g., miR-29, miR-200c, and miR-31) and are involved in the osteogenic differentiation and wound regeneration [6, 7]. On the other hand, transfection of cells with any nucleic acids, including miRs, is always a challenge, due to limitations on the cellular delivery. Researchers use various approaches to overcome these limitations by using transfecting agents, which include viruses, lipids, polymers, and peptides [8]. These transfecting agents operate as carriers to deliver the nucleic acids through the cell membranes into the cells. Among them, nonviral cationic biomaterials, like polymers or lipids, have attracted significant attention among researchers due to the versatility, enhanced safety, and low toxicity and immunogenicity [9].

In this chapter, we present the necessary steps to perform wound repair assay or scratch assay and all the critical insights required for the successful completion of the study. We also present the case of how nucleic acid transfection, such as using miRs, can affect the scratch-healing properties of a cell line.

2 Materials

2.1 Reagents

1. Medium: The media used will depend on the cultured cell line (*see* **Note 1**) and its requirements for growth. For example, J774 and Raw264 murine macrophages require Dulbecco's modified Eagle's medium (DMEM) supplemented with 10% fetal bovine serum (FBS) and 1% penicillin–streptomycin antibiotic.
2. 1× 0.25% trypsin with 0.1% EDTA (to enhance the efficacy of trypsin) in Hank's Balanced Salt Solution (HBSS) without calcium, magnesium, and sodium bicarbonate, if the cell line requires trypsinization for cell harvesting.
3. Sterile 1× phosphate-buffered saline (PBS).
4. 50 μg/ml of poly-L-lysine freshly prepared in 1× PBS.
5. 0.5 mg/ml of mitomycin C stock solution in distilled water to reduce the cell proliferation. Alternatively, it is recommended to use reduced serum media during the incubation for different time points to prevent cell proliferation.
6. Reduced serum media (i.e., OPTIMEM) can be purchased from supplier; otherwise, regular cell culture media without

adding any serum or antibiotic can be used as reduced serum media.

7. Transfecting agent, such as Lipofectamine 2000, to transfect the nucleic acids inside cells.
8. Nucleic acids (e.g., plasmid or microRNA) isolated from cells or purchased from a supplier. Plasmids can also be constructed in lab with recombinant DNA technology.
9. Continuous supply of 5% CO_2 with humidity, if the cell culturing conditions require a 5% CO_2 environment for cell growth.

2.2 Equipment

1. Tissue culture incubator maintained at 37 ° C and constant supply of 5% CO_2 with humidity to prevent drying up of media.
2. 24-well tissue culture plates.
3. Sharp and narrow pipet tips for seeding and treating cells and to generate thin scratches on monolayer cells.
4. Hemocytometer to count cells for seeding.
5. Bright-field/phase-contrast inverted microscope with camera for imaging.
6. Image J image analysis software (https://imagej.nih.gov).

3 Methods

3.1 Procedure

1. If the plate is not precoated, it is necessary to coat the 24-well plate with poly-L-lysine, which will help cells to enhance the attachment of the cells in the bottom layer of the plate. Use 500 μl of poly-L-lysine with concentration of 50 μg/ml. Incubate plate at 4 °C for overnight or for 2 h at 37 °C.
2. On the following day, remove the coating material and gently wash (*see* **Note 3**) with 1× PBS to remove unbound coating substrate. It is optional to block the plate with 500 μl of 2 mg/ml bovine serum albumin (BSA) for 1 h at 37 °C. Then, wash (*see* **Note 3**) the plate with 1× PBS to remove BSA residue.
3. Harvest and seed the appropriate number of cells in each well to prepare a monolayer of cells for the scratch assay. The number of cells will vary depending on the cell lines. If the cells doubling time is less or approximately equal to 22 h, seed 0.5×10^5 cells/well. If the doubling time is slower, for example, more than 34 h, then seed 1×10^5 cells/well. Seed cells with appropriate complete media (like DMEM or F12K). Complete media can be prepared by addition of 10% FBS and 1% penicillin–streptomycin in purchased media.
4. Incubate overnight at 37 °C with continuous supply of 5% CO_2 and humidity (if applicable).

5. Next day, replace the medium with 500 μl of reduced/no serum media containing mitomycin C (10 μg/ml). Using low or no growth medium minimizes the cell proliferation, as cell proliferation can mislead the cellular migration data.
6. Incubate overnight as above.
7. Next day, transfect cells with microRNA or any other nucleic acids with the transfecting agent, such as Lipofectamine 2000, following manufacturer's protocol. The concentration of the nucleic acid will depend on the experimental and therapeutic conditions, as independently identified by other experiments, such as dose–response studies, cytotoxicity, and gene expression. Always use appropriate controls, such as scramble siRNA, for comparison. In some cases, transfection can take place using incomplete media, with or without optimum, for 6 h. Subsequently to 6 h of incubation, the wells can be washed (*see* **Note 3**) with 1× PBS and supplemented with 500 μl of complete media.
8. Incubate at 37 °C as above.
9. Next day, take a thin and narrow-mouthed 200-μl pipette tip (*see* **Note 2**) to make a horizontal scratch in each well. Here, it is easy to place the lid edge of the plate in middle of each well, acting as a guide to scratch with the tip a straight line. This way, all wells will have straight scratches horizontally to the plate, and this will assist in imaging under the microscope more easily.
10. Wash each well with 1× PBS and add 500-μl complete medium (*see* **Note 3**).
11. Take picture of each well. This image will correspond to the 0-h time point. It is necessary to identify and use a single spot in each well, for taking a picture at the different time points. Suggestion: If an inverted microscope with an electronically controlled stage is not available to identify and relocate pre-mapped locations on the plates, after placing the plate on the inverted microscope, place one tape parallel to the mechanical stage of the microscope just above of and adjacent to the plate and one tape at bottom and adjacent to the plate. Then, place the first well above the objective and locate the scratch and the spot you wish to use for all imaging in that well. Then, make a mark on the top and bottom tape, indicating the exact location of the plate's position, adding also a mark on the plate itself. Thinner and accurate pens will improve the accuracy. Repeat this for each well. Thus, for each following time point, align the top tape's mark with top plate's mark and bottom tape's mark with bottom plate's mark, to place the plate exactly at the same spot every time (Fig. 1). Repeat as needed for each well.
12. Take pictures similarly at 24 h, 48 h, and 72 h time points (*see* **Notes 4, 5** and **6**). Additional time points before, between, or

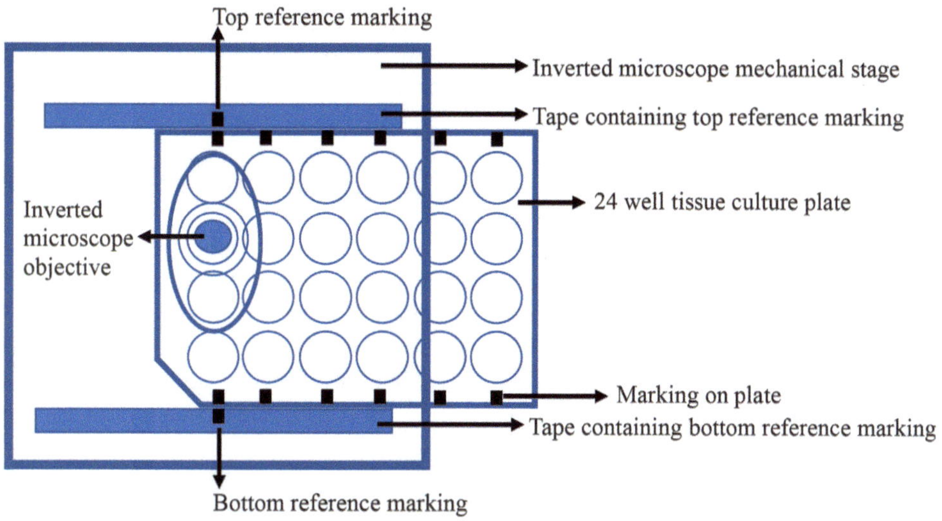

Fig. 1 Placemat of 24-well plate on inverted microscope plate to detect the same spot every time taking image

after those specified above may be required, depending on the expected experimental outcome. It is necessary to maintain the zoom and light intensity same for every time point to minimize variability in software analysis.

3.2 Data Analysis

Images from different time points can be analyzed with a variety of software. A commonly and free used software for image analysis in biological applications is ImageJ. This software can be easily download from https://imagej.nih.gov/ere. There is a plethora of available plug-ins that enhance the software's capabilities, depending on the needed application.

1. First, import each image in the ImageJ software and convert them to the TIFF format. This will help to keep the image settings among the different images with minimal variations. There are several macros (preset methods) available online, which can be used to adjust and analyze all images in a similar manner. However, it is preferable to prepare your own macros, which will permit to adjust the image easily and accurately based on the individual microscope's light intensity and zoom. To do this, open the image in ImageJ and then from the menu bar select plugins, then select macros, and then select record.
2. From the menu bar of ImageJ software, select "Type" tab and then select subtab "16-bit."
3. Go to "Process" tab at the menu bar and then select "Find Edges."
4. Select the "Image" tab from the menu bar and select "Sharpen."

5. Form the same tab, select "Adjust" and select subtab "Threshold." A new window will appear, and select the Default mode and B&W mode from the dropdown menu and select apply.
6. Repeat **step 3**.
7. Go to "Image" tab and select "Lookup Tables" and then select subtab "Invert LUT."
8. Finally, select the "Analyze" tab and select "Analyze Particles." A new window will appear, where you should select the "Summarize" option to generate data. In this stage, we can export image as JPEG format and data in a tabular format. The data will represent the covered and noncovered area of the image by the cells, which can be used to quantitatively evaluate the gap closure as a function of time.

4 Notes

1. Selection of cell line while doing this experiment is very important. Some cells, like immune cells, do no attach in the bottom of the plate. So, it is important to select cells that easily attach in the bottom of the plate.
2. While making the scratch, it is good to use 200-μl tip with sharp nozzle which allows to make straight scratch in the plate. Additionally, while making a scratch, it is also helpful to make a straight line in a white paper and then place the plate on top and then make a scratch.
3. During the washing step, it is necessary to be careful and add 1× PBS slowly, up against the walls of the wells. Forceful addition of PBS can disrupt the scratch layer and cells can be removed.
4. While doing the experiment for longer time points, such as 96 h, it is good to replace the media after 72 h.
5. While the taking pictures under a microscope, it is very important to keep the lens and zoom identical for each time point to diminish any ambiguity. Additionally, light exposure is also very important. Excess or dim light can cause blur images during the analysis.
6. While handling the plate (during taking the pictures), it is very important not to open the lid, as this may cause contamination and can compromise the experiment.

References

1. Liang CC, Park AY, Guan JL (2007) In vitro scratch assay: a convenient and inexpensive method for analysis of cell migration in vitro. Nat Protoc 2(2):329–333. https://doi.org/10.1038/nprot.2007.30
2. Gurtner GC, Werner S, Barrandon Y et al (2008) Wound repair and regeneration. Nature 453 (7193):314–321. https://doi.org/10.1038/nature07039

3. Takeo M, Lee W, Ito M (2015) Wound healing and skin regeneration. Cold Spring Harb Perspect Med 5(1):a023267. https://doi.org/10.1101/cshperspect.a023267
4. Main KA, Mikelis CM, Doci CL (2019) In vitro wound healing assays to investigate epidermal migration. Methods Mol Biol 2109:147–154. https://doi.org/10.1007/7651_2019_235
5. Hossian A, Sajib MS, Tullar PE et al (2018) Multipronged activity of combinatorial miR-143 and miR-506 inhibits lung cancer cell cycle progression and angiogenesis in vitro. Sci Rep 8(1):10495. https://doi.org/10.1038/s41598-018-28872-2
6. Zhang Y, Ma W, Zhan Y et al (2018) Nucleic acids and analogs for bone regeneration. Bone Res 6:37. https://doi.org/10.1038/s41413-018-0042-7
7. Aunin E, Broadley D, Ahmed MI et al (2017) Exploring a role for regulatory miRNAs in wound healing during ageing: involvement of miR-200c in wound repair. Sci Rep 7(1):3257. https://doi.org/10.1038/s41598-017-03331-6
8. Ni R, Feng R, Chau Y (2019) Synthetic approaches for nucleic acid delivery: choosing the right carriers. Life (Basel) 9(3). https://doi.org/10.3390/life9030059
9. Labatut AE, Mattheolabakis G (2018) Non-viral based miR delivery and recent developments. Eur J Pharm Biopharm 128:82–90. https://doi.org/10.1016/j.ejpb.2018.04.018

Chapter 9

In Vivo Ear Sponge Lymphangiogenesis Assay

Racheal G. Akwii, Md S. Sajib, Fatema T. Zahra, Hanumantha R. Madala, Kalkunte S. Srivenugopal, and Constantinos M. Mikelis

Abstract

Lymphangiogenesis, the formation of lymphatic vessels from preexisting ones, is an important process in wound-healing physiology. Deregulation of lymphangiogenesis and lymphatic vascular remodeling have been implicated in a range of inflammatory conditions, such as lymphedema, lymphadenopathy, tumor growth, and cancer metastasis. Any attempt in understanding various parameters of the lymphangiogenic process and developing desirable therapeutic targets requires recapitulating these conditions in in vivo models. One pitfall with some experimental models is the absence of immune response, an important regulatory factor for lymphangiogenesis. We overcome this issue by using immune competent mice. In this chapter, by using Angiopoietin-2 (Ang2), a protein that belongs to the Ang/Tie signaling pathway, we describe the ear sponge assay with important adaptations, highlighting a reproducible and quantitative tool for assessment of in vivo lymphangiogenesis.

Key words Lymphangiogenesis, In vivo, Ang2, Lyve-1, Ear sponge

1 Introduction

The lymphatic system is an integral part of the circulatory system, containing the lymphatic vessels, lymph nodes, tonsils, adenoids, spleen, and thymus. Its main function is to transport lymph, a fluid containing white blood cells, toxins, and cellular waste, throughout the body. Lymphangiogenesis is the formation of lymphatic vessels, which is synonymous to angiogenesis, the formation of blood vessels [1]. Lymphangiogenesis is an important process in human physiology and pathophysiology of certain diseases, given the critical function of the lymphatic vessels in regulating tissue homeostasis, immune cell trafficking, absorption of dietary fats and, through facilitating the flow of lymph back to the blood, maintaining normal blood volume and pressure, preventing edema [2, 3]. The role of lymphangiogenesis was until recently overlooked in disease pathology; however, the identification of lymphatic endothelial markers in the past few years has advanced our knowledge about

Hiranmoy Das (ed.), *Wound Regeneration: Methods and Protocols*, Methods in Molecular Biology, vol. 2193,
https://doi.org/10.1007/978-1-0716-0845-6_9,

the molecular pathways regulating lymphangiogenesis. The common diseases associated with dysfunctionality of the lymphatic system are lymphadenopathy (enlargement of the lymph nodes) [4–6], lymphedema (swelling due to lymph node blockage) [7–9], and associated cancers [10–12].

The development of the skin lymphatic system was described more than 100 years ago in pigs [13]. The primitive plexus of lymphatic vessels forms in the border between dermis and subcutaneous tissue, and later during development, the lymphatic plexus invades the dermis, giving rise to a secondary plexus, with the initial one functioning as a collective plexus [13]. The adult skin plexus can be distinguished in the superficial and deep lymphatic plexus. The superficial plexus extends into the dermal papillae, and the deep is located in the lower dermis [14]. The anatomy of the mouse ear skin offers an ideal model for in vivo assessment of lymphatic vascular functions and, thus, is preferred for lymphangiogenesis models [15].

Angiopoietin-2 (Ang2) is one of the proteins of the Ang/Tie signaling pathway [16] with a demonstrated role in lymphangiogenesis regulation [17–20]. Here, we use Ang2 as a lymphangiogenesis inducer in the ear sponge assay. The technique involves a superficial incision on the dorsal skin area of the mouse ear, creating a pocket between the external superficial ear surface and cartilage where the collagen-coated sponge is embedded. This technique, which is also used for angiogenesis evaluation, is cost effective, easy to perform, and versatile, as different growth factors, inhibitors, or cancer cells can be used and techniques, such as immunofluorescence, immunohistochemistry, or bioluminescence, can be applied [21, 22]. In this chapter, we will further focus on the immunofluorescence analysis of the sponges and quantification of the lymphatic vessels.

2 Materials

Prepare all solutions using ultrapure water and analytical grade reagents. The preparation of the reagents should take place on ice under sterile conditions, unless specified. Amounts can vary depending on the required total volume.

1. Collagen type 1 (*see* **Note 1**).
2. 1× phosphate-buffered saline (PBS): Add NaCl 8 g/l, KCl 0.2 g/l, Na_2HPO_4 1.44 g/l, KH_2PO_4 0.24 g/l in distilled water. Adjust pH to 7.4. Prior to use, the buffer is autoclaved for 20 min at 121 °C.

3. 1 M HEPES: For 0.5 l solution, add 119.15 g of HEPES and dissolve in 400 ml of PBS. Adjust the final volume to 500 ml with PBS.
4. 0.2 M NaOH: For 0.5 l solution, add 4 g of NaOH and dissolve in 400 ml of sterile H_2O. Adjust volume to 500 ml with sterile H_2O.
5. 0.1% acetic acid: For 0.5 l solution, add 0.5 ml of acetic acid 99% solution in 500 ml of sterile H_2O.
6. 3% bovine serum albumin (BSA) in PBS: For 10 ml solution, dissolve 0.3 g BSA and in 8 ml 1 × PBS. Adjust the final volume to 10 ml with PBS and filter sterilize.
7. PBS containing 0.2% Triton X-100 and 200 mM glycine: For 50 ml solution, dissolve 751 mg glycine in 40 ml PBS and vortex to dissolve. Add PBS to reach 50 ml final volume and add 100 μl Triton X-100 and vortex to dissolve.
8. Starvation medium with phenol red to act as an indicator when neutralizing the pH of the collagen mix (if not neutralized, the acetic acid may cause inflammation to the mouse ear).
9. Murine recombinant Ang2 (or other stimulator).
10. Syringe filters with 0.2 μm pore size.
11. Syringes.
12. 1.5-ml and 2-ml microcentrifuge tubes.
13. 96-well plates.
14. Polyclonal goat anti-mouse LYVE-1 antibody (R&D Systems, Cat No: AF2125).
15. Anti-goat Alexa Fluor 488 antibody.
16. Hoechst.
17. Sharp tip dissecting forceps.
18. Surgical scissors.
19. Sterile nylon nonabsorbable suture with 3/8 needle or smaller.
20. Sterile 3-mm biopsy punch.
21. Sterile absorbable gelatin sponge 12–7 mm.
22. Mounting media.
23. Sterile tips and pipettes.
24. Isoflurane.
25. 70% ethanol.
26. Warming plate.
27. OCT.
28. 4% paraformaldehyde.
29. ImageJ software (*see* **Note 2**).

3 Methods

Diligently follow all waste disposal regulations when disposing waste materials. Animal experiments should take place according to the Institutional Animal Care and Use Committee (IACUC)-approved protocols of the host institution, in compliance with the Guide for the Care and Use of Laboratory Animals.

3.1 Sponge Coating

3.1.1 Preparation of Collagen Mix

Work in a sterile environment (i.e., laminar flow hood). Clean the work surface with 70% ethanol. Prepare the mixture on ice.

1. In an appropriate tube, add the calculated volume of the 0.1% acetic acid, collagen type I stock solution, and phenol red-containing media, following the ratio: 4:4:1 for acetic acid: collagen: medium. The color of the mixed solution should be yellow.
2. Add 0.2 M NaOH dropwise, while flicking the tube to mix. The solution should turn salmon pink. Embed the tube periodically on ice to maintain low temperature.
3. Continue adding 0.2 M NaOH dropwise, while flicking the tube, until the color solution turns pink.
4. Add 1 M HEPES dropwise until the pink solution turns salmon pink again.
5. Keep the mixture on ice.

3.1.2 Sponge Preparation

Work in a sterile environment (i.e., laminar flow hood). Clean the work surface with 70% ethanol.

1. Resuspend the lymphangiogenic agent in starvation medium or PBS at the desired concentration. Calculate the final quantity in 20 μl of medium that corresponds to each sponge. Calculate all reagents for one extra sponge than what is experimentally planned.
2. Using a sterile biopsy punch, cut the gelatin sponge into cylindrical pieces of 3 mm^3. Cut as many as required, depending on the groups and/or replicates of the experiment.
3. Using sterile forceps, carefully place sponges in a 96-well plate, one per well.
4. Pipette 20 μl of medium (for control group) and agent in medium (for treatment group) into the corresponding sponges.
5. Incubate for 30 min at 37 °C, which allows the sponge to absorb the medium.
6. Coat the sponges with collagen by dipping each sponge in the collagen mix three times using sterile forceps (*see* **Note 3**).

7. Remove the sponges from collagen and place them back into the 96-well plate, one per well.
8. Incubate for 30 min at 37 °C.

3.2 Sponge Implantation

Work in a sterile and disinfected environment. Keep all the equipment disinfected using 70% ethanol or autoclaved in 121 °C for 20 min. Male or female mice 8- to 12-week-old are used.

1. Anesthetize mice with isoflurane inhalation using an anesthesia induction chamber (0.5–2% isoflurane, flow: 1 l/min).
2. Wait for 5 min until the mouse is anesthetized.
3. Confirm anesthesia using the toe pinch reflex.
4. Lay the mouse in a prone position on the warming plate, keeping them under continuous anesthesia.
5. Wipe the outer ear lobes using 70% ethanol. Alcohol swabs can be used.
6. Make a small superficial horizontal incision in the dorsal (external) surface of the mouse ear skin (*see* **Note 4**).
7. Smoothly and gradually, detach the external mouse ear skin layer from the cartilage, using thin sterile forceps (*see* **Note 5**). The size of the pocket of the detached skin should exceed the size of the sponge, so that it can be easily embedded.
8. Using forceps, place the gelatin sponge inside the pocket of the detached skin. Use the forceps smoothly to secure the full insertion of the entire sponge in the pocket, with the minimum possible manipulation and deformation of the sponge.
9. Make a suture point in the middle of the incision to join the dorsal skin of the two sides of the incision (*see* **Note 6**).
10. Sponge can remain implanted for 14 or 21 days (*see* **Note** 7).

3.3 Sponge Processing

1. After 14 or 21 days, euthanize the mice in accordance to the guidelines of the IACUC-approved protocols of the host institution, in compliance with the Guide for the Care and Use of Laboratory Animals.
2. Remove the ear area containing the sponge (*see* **Note 8**).
3. On a dissection board, using forceps, gently separate the ear skin layers and then cut the surrounding tissue.
4. Harvest sponges and embed them in tissue OCT (*see* **Note 9**).
5. Keep OCT-embedded samples on dry ice until OCT solidifies (store in −80 °C or proceed to next step).
6. Perform cryosections of 10 μm thickness. Keep the sections on dry ice and then store the sections in −80 °C or proceed with the staining.

7. Fix sections in 4% paraformaldehyde for 20 min at room temperature (RT). Wash with PBS (twice × 5 min).
8. Permeabilize for 15 min with 0.2% Triton X-100 + 200 mM glycine at RT. Wash with PBS (twice × 5 min).
9. Block sections for 2 h in 3% BSA in PBS (blocking solution) at RT.
10. Pipette appropriate volume of LYVE-1 antibody to reach a concentration of 10 μg/ml in blocking solution and incubate overnight at 4 °C (*see* **Note 10**).
11. Perform two quick (1–3 min) and then two 5-min washes with PBS.
12. Incubate with appropriate Alexa Flour 488 (green) or Alexa Fluor 546 (red) secondary antibodies in a 1:200 dilution in blocking solution for 2 h at RT in the dark.
13. Perform two quick (1–3 min) and then two 5-min washes with PBS.
14. For nuclear staining, use Hoechst diluted 1:2000 in PBS for 20 min in the dark at RT.
15. Wash with PBS twice, 5 min each. Remove excess PBS.
16. Mount using mounting media and carefully cover with a coverslip.
17. Store stained slides at 4 °C.
18. Fluorescent images can be acquired with an epifluorescent or confocal microscope (Fig. 1).

3.4 Quantification of the Lymphatic Vessels

The quantification of the lymphatic vessels takes place with ImageJ software, which is freely available to download (*see* **Note 2**). The quantification can vary, depending on the magnification. Low magnification can allow more automated analysis (Thresholding), but will not allow the inclusion of the hollow cavity of large lymphatic vessels. High magnification can include this in the calculation and eliminate potential off-target fluorescence, but is performed manually (to our knowledge) and thus is more time- and effort-consuming. Below, we describe both quantification options.

3.4.1 Scale Calibration (See Note 11)

1. Open ImageJ software.
2. Insert Image, by selecting File → Open → Browse in the files and select the image or just drag and drop the image on the ImageJ window.
3. Zoom In the image (if needed), by selecting the "Magnifying glass" option.
4. Select the "Straight" line selection and "draw" a line overlying on the image scale bar (*see* **Note 12**).

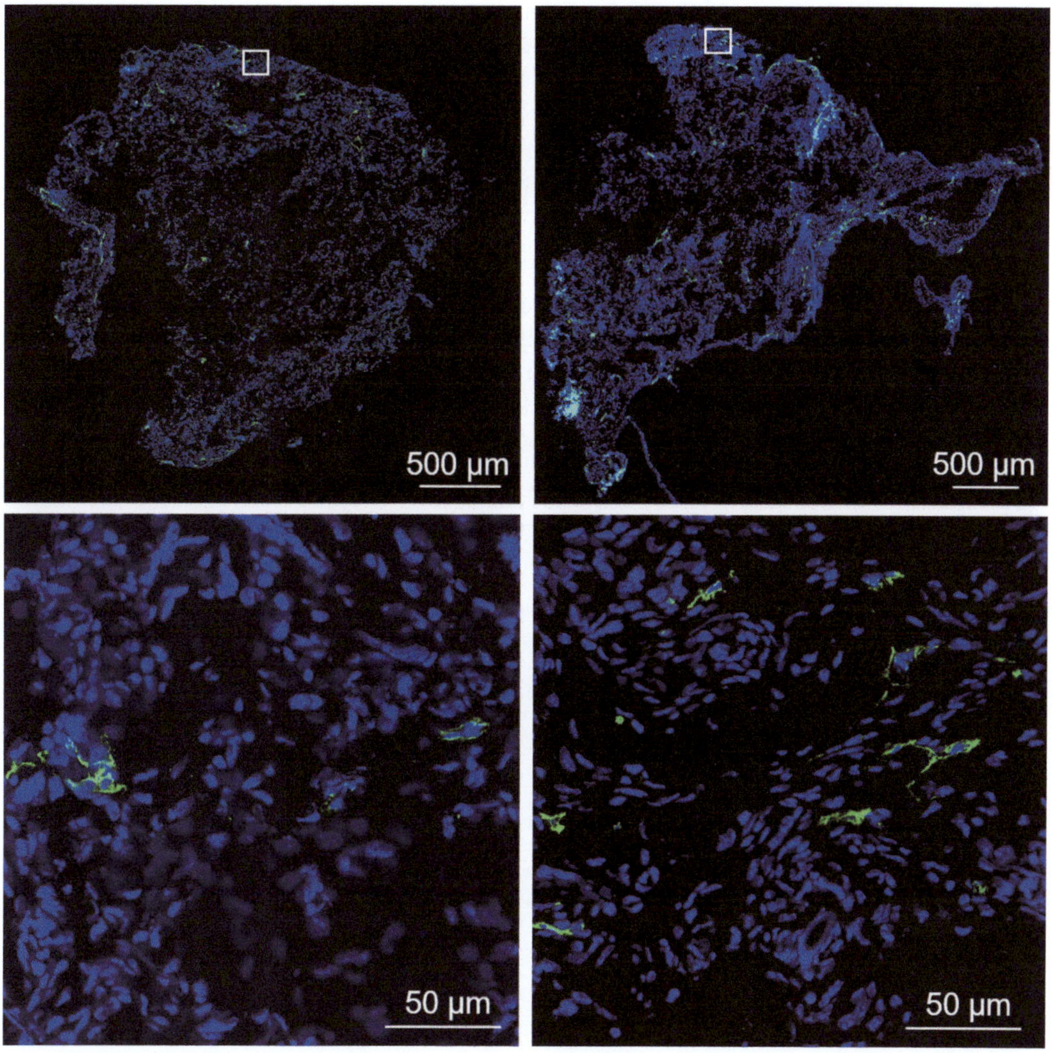

Fig. 1 Representative immunofluorescent image of sponges containing PBS (control; left panel) or stimulation (Ang2; right panel) and magnified areas (lower panels). Lymphatic vessels are identified through Lyve-1 staining (green) and cell nuclei appear in blue (Hoechst)

5. Select Analyze → Set Scale. The "Distance in pixels" window shows the distance of the line drawn in pixels. Fill the "Known distance" window with the distance that corresponds to the selected line and fill the "Unit of length" window with the appropriate length unit (i.e., nm). Press "OK."
6. Zoom Out (if needed), by selecting the "Magnifying glass" option with right click option (for right-handed users).

   After the calibration, the quantification can take place with automatic setup (thresholding; Subheading 3.4.2) or manually (Subheading 3.4.3).

3.4.2 Thresholding (See ***Note 13****)*

1. If the image contains different stainings/colors, you should isolate the color of interest (lymphatic vessel staining). Select: Image → Color → Make composite. Then select: Image → Color → Split channels.
2. Save the split color images.
3. Select and open the image with the color of interest (lymphatic staining).
4. Select Image → Adjust → Threshold. It should have as a default the Red color (for contrast from the black and white background). Adjust the parameters in such a way so that the Red will overlap the green fluorescence in our experiment (if the lymphatics are stained with red fluorescence, you can choose another color to adjust the threshold) (Fig. 2).
5. Select Apply.
6. Select Analyze → Analyze Particles. Fix the parameters (*see* **Note 14**).

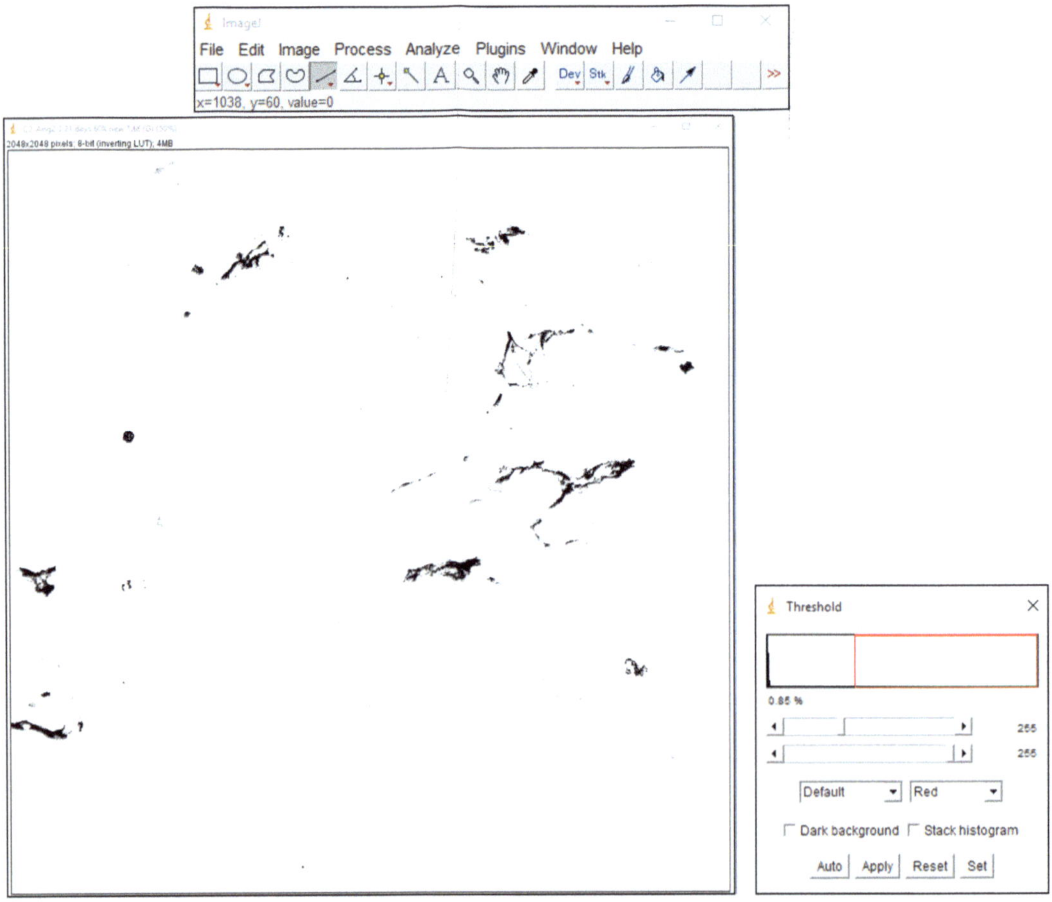

Fig. 2 Representative image of the Threshold adjustment of the magnified image from Fig. 1

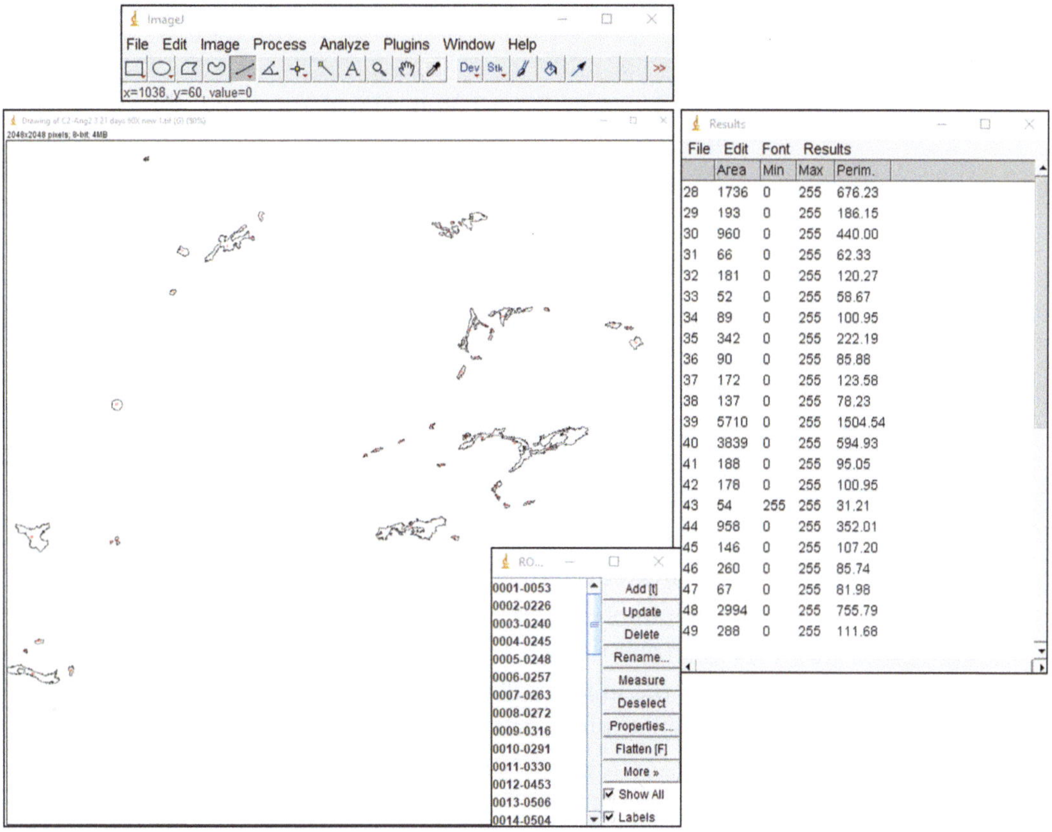

Fig. 3 Representative image of the area and perimeter quantification of the lymphatic areas in the magnified image from Fig. 1

7. Select OK.
8. You have a series (list) of results and a diagram showing each area quantified with the area and perimeter values (Fig. 3).
9. Select Results → Summarize in the results window and you will get the mean, SD, minimum, and maximum values.
10. The list can be exported to excel form and saved.

3.4.3 Manual Quantification

1. Open the image.
2. Select the "Freehand selection."
3. Move the mouse cursor to draw around each vessel.
4. Select Analyze → Measure.
5. A new window showing the result of the area quantified and the perimeter appears. New quantifications will be added there.
6. Select Results → Summarize in the results window and you will get the mean, SD, minimum, and maximum values.
7. The list can be exported to excel form and saved.

4 Notes

1. Collagen type I concentration may vary. The concentration of the commercially available high-concentration collagen stock we use is about 8.21 mg/ml, which provides a final collagen concentration in the collagen mix of 3.5–4 mg/ml.
2. ImageJ software can be downloaded free of charge from the following website: https://imagej.nih.gov/ij/download.html.
3. When coating the sponges with collagen, start by coating the control sponge and then the treated sponge to avoid mixing of the agent with the control.
4. The incision should be superficial so that it does not cut the skin of the other side. If that happens, that ear should be excluded from the study.
5. In case the thin forceps perforate the skin, that ear should be excluded from the study.
6. After implantation of the sponge, monitor the mouse recovery from anesthesia.
7. For the first few days after sponge implantation, monitor the mice for inflammation or infection.
8. If many mice are simultaneously euthanized, the ears can be placed in marked wells in a 24-well plate, with PBS on ice, until all sponges are harvested.
9. When preparing the sponges for sectioning, ensure that the sponge is fully embedded in OCT.
10. During antibody staining, keep the slides in an air-sealed container containing moist paper towel to eliminate evaporation, especially during overnight incubation.
11. Scale calibration can only take place if the image has a scale bar or the image dimensions are known.
12. By clicking "Shift" while drawing the line, the variability in the angles of the line is eliminated.
13. If the image contains letters, scale bar, or anything else that can affect your measurements, you can select a specific area of the image that you would like to analyze by the following steps: Select the "Rectangular" option, select the area you want to analyze, and select: Image → Duplicate.
14. The size of the pixel (0–infinity) should be over "0," to avoid background measurement. We normally set it up to 10 (10–infinity). In the "Show" window, we select "Outlines." The boxes "Display results," "Clear results," "Add to Manager,"

and "Include holes" are normally checked. The box "Exclude on edges" is clicked if we do not want to calculate the vessels in the edge of the image.

Acknowledgments

This work was supported in part by grants from the National Institutes of Health (NCI) R15CA231339, Texas Tech University Health Sciences Center (TTUHSC) Office of Research to C.M.M, and Cancer Prevention and Research Institute of Texas (CPRIT, RP170207) to K.S.S. The funders had no role in study design, decision to write, and preparation of the manuscript.

References

1. Van de Velde M, Garcia-Caballero M, Durre T et al (2018) Ear sponge assay: a method to investigate angiogenesis and lymphangiogenesis in mice. Methods Mol Biol 1731:223–233. https://doi.org/10.1007/978-1-4939-7595-2_20
2. Tammela T, Alitalo K (2010) Lymphangiogenesis: molecular mechanisms and future promise. Cell 140(4):460–476. https://doi.org/10.1016/j.cell.2010.01.045
3. Baker A, Kim H, Semple JL et al (2010) Experimental assessment of pro-lymphangiogenic growth factors in the treatment of post-surgical lymphedema following lymphadenectomy. Breast Cancer Res 12(5):R70. https://doi.org/10.1186/bcr2638
4. Geladari E, Dimopoulou G, Margellou E et al (2019) Coexistence of Hodgkin and non-Hodgkin lymphoma; composite lymphoma (CL) in a patient presenting with waxing and waning lymphadenopathy. Cardiovasc Hematol Disord Drug Targets. https://doi.org/10.2174/1871529x19666191014111118
5. Gaddey HL, Riegel AM (2016) Unexplained lymphadenopathy: evaluation and differential diagnosis. Am Fam Physician 94(11):896–903
6. Otsuka I (2019) Cutaneous metastasis after surgery, injury, lymphadenopathy, and peritonitis: possible mechanisms. Int J Mol Sci 20 (13). https://doi.org/10.3390/ijms20133286
7. Depairon M, Lessert C, Tomson D et al (2017) Primary lymphedema. Rev Med Suisse 13 (586):2124–2128
8. Bernas M, Thiadens SRJ, Smoot B et al (2018) Lymphedema following cancer therapy: overview and options. Clin Exp Metastasis 35 (5-6):547–551. https://doi.org/10.1007/s10585-018-9899-5
9. Agharbi FZ (2018) Lymphedema complicated with verrucous papillomatosis. Pan Afr Med J 31:251. https://doi.org/10.11604/pamj.2018.31.251.16166
10. Gombos Z, Xu X, Chu CS et al (2005) Peritumoral lymphatic vessel density and vascular endothelial growth factor C expression in early-stage squamous cell carcinoma of the uterine cervix. Clin Cancer Res 11 (23):8364–8371. https://doi.org/10.1158/1078-0432.Ccr-05-1238
11. Liang P, Hong JW, Ubukata H et al (2006) Increased density and diameter of lymphatic microvessels correlate with lymph node metastasis in early stage invasive colorectal carcinoma. Virchows Arch 448(5):570–575. https://doi.org/10.1007/s00428-006-0166-9
12. Roma AA, Magi-Galluzzi C, Kral MA et al (2006) Peritumoral lymphatic invasion is associated with regional lymph node metastases in prostate adenocarcinoma. Mod Pathol 19 (3):392–398. https://doi.org/10.1038/modpathol.3800546
13. Sabin FR (1904) On the development of the superficial lymphatics in the skin of the pig. Am J Anat 3:183–195
14. Skobe M, Detmar M (2000) Structure, function, and molecular control of the skin lymphatic system. J Investig Dermatol Symp Proc 5(1):14–19. https://doi.org/10.1046/j.1087-0024.2000.00001.x
15. Kilarski WW, Guc E, Swartz MA (2018) Dorsal ear skin window for Intravital imaging and functional analysis of Lymphangiogenesis.

Methods Mol Biol 1846:261–277. https://doi.org/10.1007/978-1-4939-8712-2_17

16. Davis S, Aldrich TH, Jones PF et al (1996) Isolation of angiopoietin-1, a ligand for the TIE2 receptor, by secretion-trap expression cloning. Cell 87(7):1161–1169. https://doi.org/10.1016/s0092-8674(00)81812-7
17. Gale NW, Thurston G, Hackett SF et al (2002) Angiopoietin-2 is required for postnatal angiogenesis and lymphatic patterning, and only the latter role is rescued by Angiopoietin-1. Dev Cell 3(3):411–423
18. Akwii RG, Sajib MS, Zahra FT et al (2019) Role of Angiopoietin-2 in vascular physiology and pathophysiology. Cell 8(5):471. https://doi.org/10.3390/cells8050471
19. Yan ZX, Jiang ZH, Liu NF (2012) Angiopoietin-2 promotes inflammatory lymphangiogenesis and its effect can be blocked by the specific inhibitor L1-10. Am J Physiol Heart Circ Physiol 302(1):H215–H223. https://doi.org/10.1152/ajpheart.00895.2011
20. Dellinger M, Hunter R, Bernas M et al (2008) Defective remodeling and maturation of the lymphatic vasculature in Angiopoietin-2-deficient mice. Dev Biol 319(2):309–320. https://doi.org/10.1016/j.ydbio.2008.04.024
21. Garcia-Caballero M, Van de Velde M, Blacher S et al (2017) Modeling pre-metastatic lymphvascular niche in the mouse ear sponge assay. Sci Rep 7:41494. https://doi.org/10.1038/srep41494
22. Durre T, Morfoisse F, Erpicum C et al (2018) uPARAP/Endo180 receptor is a gatekeeper of VEGFR-2/VEGFR-3 heterodimerisation during pathological lymphangiogenesis. Nat Commun 9(1):5178. https://doi.org/10.1038/s41467-018-07514-1

Chapter 10

Identification of Rho GEF and RhoA Activation by Pull-Down Assays

Md S. Sajib, Fatema T. Zahra, Racheal G. Akwii, and Constantinos M. Mikelis

Abstract

The small GTPase RhoA participates in actin and microtubule machinery, cell migration and invasion, gene expression, vesicular trafficking and cell cycle, and its dysregulation is a determining factor in many pathological conditions. Similar to other Rho GTPases, RhoA is a key component of the wound-healing process, regulating the activity of different participating cell types. RhoA gets activated upon binding to guanine nucleotide exchange factors (GEFs), which catalyze the exchange of guanosine diphosphate (GDP) for guanosine triphosphate (GTP). GTPase-activating proteins (GAPs) mediate the exchange of GTP to GDP, inactivating RhoA, whereas guanine nucleotide dissociation inhibitors (GDIs) preserve the inactive pool of RhoA proteins in the cytosol. RhoA and Rho GEF activation is detected by protein pull-down assays, which use chimeric proteins with Rhotekin and G17A mutant RhoA as "bait" to pull down active RhoA and RhoA GEFs, respectively. In this chapter, we describe an optimized protocol for performing RhoA and GEF pull-down assays.

Key words RhoA, RhoA pull-down assay, Rhotekin, GEF, GEF pull-down assay

1 Introduction

The family of Rho GTPases belongs to the Ras superfamily of small GTP-binding proteins and its members are expressed in all eukaryotes. The best-characterized Rho GTPases are Ras homolog gene family, member A (RhoA), Ras-related C3 botulinum toxin substrate 1 (Rac1) and cell division control protein 42 (Cdc42). The small GTPase RhoA is a key regulator of cytoskeletal proteins, regulating from cellular morphology and cytokinesis to proliferation and transcriptional activity [1–3], and its role has been demonstrated during the wound-healing process [4, 5]. Like most G proteins, RhoA cycles between active GTP-bound and inactive GDP-bound states. When inactive, RhoA is found in the cytoplasm bound to membranes of subcellular organelles or associated with GDIs, but upon GTP binding it gets tethered to the cell

Hiranmoy Das (ed.), *Wound Regeneration: Methods and Protocols*, Methods in Molecular Biology, vol. 2193, https://doi.org/10.1007/978-1-0716-0845-6_10,

membrane, where it activates its effectors [6–9]. Several proteins, such as Rhotekin, Rhophilin, and Protein Kinase N, bind specifically to active RhoA and inhibit its intrinsic GTPase activity. The Rho-binding domain of Rhotekin strongly interacts with RhoA, but not with Rac1 or Cdc42 [10, 11]. For the RhoA pull-down assay, the N-terminal part of Rhotekin is expressed as a chimeric protein with glutathione-S-transferase and is coupled to glutathione beads to pull down active RhoA from cell lysates [12].

Approximately 80 GEFs have been reported in human cells. Most of the Rho GEFs have a Dbl homology (DH) domain, often followed by pleckstrin homology (PH) domain. The DH domain is responsible for the GDP exchange for GTP of Rho GTPases [13 15]. Some GEFs activate more than one small GTPases and others are specific for one, while multiple GEFs can activate the same GTPase [16]. The GEFs facilitate their exchange activity by disrupting the tight nucleotide binding. While the nucleotide binding site is not directly affected, GEFs promote nucleotide ejection by altering the conformation of the switch regions, disrupting the coordination of the Mg^{2+} ion [17]. The Rho GTPase family shares structural similarity with Ras; therefore, research on Ras structure expedited the knowledge for the activation sites of the Rho GTPase family. The dominant negative S17N-Ras mutant was shown to bind GDP with similar efficiency with the wild-type conformation, but the binding efficiency with GTP was abrogated, providing an ideal binding target for GEF binding [18]. In a similar way, the dominant-negative G17A RhoA mutant lacks Glycine at residue 17 that is critical for GTP binding and serves as bait for active GEF binding [19, 20].

The coordinated balance of GEFs and GAPs defines the final activation of each small GTPase. Dysregulation of RhoA signaling in different cell types has been implicated in a wide variety of vascular and inflammatory diseases, gastrointestinal tract disorders, and cancer [21–24]. The optimized protocol described here can be used for efficient assessment of RhoA and GEF activity in cellular or tissue samples.

2 Materials

Prepare all solutions using ultrapure water and analytical-grade reagents. The preparation of the reagents should take place on ice, unless specified. Amounts can vary depending on the required total volume.

2.1 Disposables and Reagents

1. GST-RBD Plasmid (Cat No: 15247, Addgene) (*see* **Note 1**) [12].
2. GST-G17A Plasmid (Cat No: 69357, Addgene) (*see* **Note 2**) [20].

3. Competent *E. coli* bacteria, BL21(DE3) strain.
4. Tryptone.
5. NaCl.
6. KCl.
7. Yeast extract.
8. Agar.
9. NaOH.
10. pH test strips.
11. Ampicillin hydrochloride (*see* **Note 3**).
12. Chloramphenicol (*see* **Note 4**).
13. 70% Ethanol.
14. Isopropanol anhydrous.
15. Syringe filters with 0.2 μm pores.
16. 10- and 20-mL Syringes.
17. Super optimal broth with catabolite repression medium (SOC medium; *see* **Note 5**).
18. Isopropyl-β-D-thiogalactoside (IPTG; *see* **Note 6**).
19. Glutathione Sepharose® 4B beads.
20. Aprotinin (*see* **Note 7**).
21. Leupeptin (*see* **Note 8**).
22. Dithiothreitol (DTT; *see* **Note 9**).
23. Phenylmethanesulfonyl fluoride (PMSF, *see* **Note 10**).
24. Sodium orthovanadate (Na_3VO_4; *see* **Note 11**).
25. HEPES.
26. Triton X-100.
27. β-glycerol phosphate,
28. Ethylenebis(oxyethylenenitrilo)tetraacetic acid (EGTA).
29. Magnesium chloride hexahydrate.
30. Protein estimation kit.
31. 2× and 5× SDS-PAGE sample buffer.
32. Eukaryotic cells in culture.
33. Basal cell culture medium (depending on the cell line).
34. Fetal bovine serum (FBS; *see* **Note 12**).
35. Pen/Strep (Antibiotic cocktail).
36. Phosphate-buffered saline (PBS), Sterile (*see* **Note 13**).
37. Glycerol.
38. Parafilm.

39. Disposable 15- and 50-mL conical centrifuge tubes.
40. Erlenmeyer flasks.
41. 1.5-mL microcentrifuge tubes
42. Sterile tips and micropipettes.
43. Polystyrene petri dishes, sterile.
44. Bacterial cell spreaders.
45. 6-well cell culture plates
46. Disposable cell scraper.

2.2 Equipment

1. Autoclave.
2. Incubator and shaker for bacteria.
3. Water bath.
4. Tube rotator.
5. Probe sonicator.
6. Centrifuge with temperature control.
7. Standard cell culture equipment.

3 Methods

3.1 Preparation of Luria Bertani (LB) Medium and Agar Plates

1. For 1 L of LB medium, weigh 10 g Tryptone, 10 g NaCl, and 5 g Yeast extract and dissolve in 900 mL of purified H_2O.
2. When completely dissolved, adjust the pH to 7.5 using pH test strips with NaOH solution (1 N).
3. Adjust the final volume to 1 L with purified H_2O.
4. Distribute 500 mL of the prepared LB medium into two Erlenmeyer flasks, so that each contains 250 mL. To prepare agar plates, add agar at a concentration of 15 g/L of LB medium. Add the agar and the appropriate LB medium volume in a separate flask.
5. Sterilize the solutions at 121 °C for 20 min.
6. Let the solutions come to room temperature (RT). Add ampicillin (100 μg/mL) and chloramphenicol (34 μg/mL) to the autoclaved solutions. For the agar-containing solution, addition of antibiotics should take place while the solution is tolerably warm to hold (>40 °C) before the solution solidifies (*see* **Note 14**).
7. For the agar-containing solution, once the antibiotic(s) are added, pour the solution to several petri dishes (approx. 15–20 mL/10 cm petri dish). These petri dishes should be left at room temperature until their content solidifies. Then, they can be either inverted and stored at 4 °C or used immediately (*see* **Note 15**).

3.2 Preparation of Rhotekin/GEF Beads

Day 1: Transformation of Plasmid into BL21 Competent Bacteria

1. Thaw BL21 bacteria and plasmid on ice.
2. Warm the agar plates in the incubator and set the water bath to 42 °C. Transfer 50 μL of bacteria in a sterile microcentrifuge vial and add ~1 μg of appropriate plasmid (Addgene#15247 or Addgene#69357). Mix by flicking gently (~10 s) and allow the mix to rest on ice for 30 min.
3. Perform the heat shock step by immersing the microcentrifuge tube in water bath at 42 °C for 50 s. Then, place it back on ice for 2 min.
4. Add 200 μL of SOC media in the microcentrifuge tube.
5. Incubate the bacteria at 37 °C for 1 h under continuous agitation.
6. After the incubation period, transfer 50 μL of the bacteria-SOC solution to an agar-covered petri dish and the remaining 200 μL to another. Spread the bacteria smoothly onto each agar-covered petri dish with a cell spreader (*see* **Note 16**).
7. Keep the plate in an upright position for 5 min to allow the bacteria to settle on the agar.
8. Incubate overnight at 37 °C in inverted position.

   Day 2: Bacterial Colony Selection and Overnight Bacterial Growth

9. Select a single, distinct colony from the agar plates and inoculate into 6–7 mL of sterile LB medium with ampicillin (100 μg/mL) and chloramphenicol (34 μg/mL).
10. Allow the liquid culture to grow overnight at 37 °C under continuous agitation.

    Day 3: Preparation of Starter Bacterial Culture

11. Take 1 mL from the best-grown overnight liquid culture and inoculate into a sterile 50 mL conical centrifuge tube containing 25 mL LB media with ampicillin (100 μg/mL) and chloramphenicol (34 μg/mL).
12. The remaining culture can be used to prepare glycerol stocks and kept at −80 °C for future use (*see* **Note 17**).
13. Prepare and autoclave 2 flasks with 250 mL LB media as described in Subheading 3.1 and let them stay at room temperature overnight. Add ampicillin and chloramphenicol. These will be used on Day 4.

    Day 4: Induction of GST-RBD/ GST-G17A Expression

14. Equally split the overnight culture (~12.5 mL each) into the previously autoclaved 250 mL LB with ampicillin (100 μg/mL) and chloramphenicol (34 μg/mL) and let the bacteria grow for 4 h (or until the OD_{600} reaches 0.6–0.8) at 37 °C under continuous agitation.

15. Add IPTG at a final concentration of 200 μM and culture overnight under continuous agitation at room temperature.

    Day 5: Bead Preparation

16. Pellet the overnight culture by centrifuging at 3000 × *g* for 30 min at 4 °C. Discard the supernatant (*see* **Note 18**).
17. Resuspend the bacteria with 12.5 mL of the following buffer: PBS containing 1% Triton X-100, 1 mM PMSF, 10 μg/mL Aprotinin, and 10 μg/mL Leupeptin. Always keep on ice (*see* **Note 19**).
18. After resuspension, transfer the lysate to pre-chilled 50 mL centrifuge tubes. Completely freeze the lysate in dry ice-ethanol mixture or liquid nitrogen. Thaw at 37 °C in H_2O. Repeat the freeze–thaw cycle three more times.
19. Sonicate the samples five times for 10 s with 10 s intervals. If the sonicator has adjustment settings, use energy setting 5 or higher (*see* **Note 20**).
20. Centrifuge the samples at 30,000 × *g* for 30 min at 4 °C.
21. Transfer the supernatant in sterile 15 mL conical centrifuge tubes and incubate it overnight at 4 °C with glutathione-sepharose beads. Add 500 μL of bead solution in 12.5 mL of lysate.

    Day 6: Preparation of Bead Aliquots and Storage

22. Centrifuge the tubes at 1,600 × *g* for 10 min at 4 °C. During this time, label the 1.5 mL microcentrifuge tubes that will store the beads and keep them on ice.
23. Carefully discard the supernatant without disturbing the beads. Add 10 mL of pre-chilled PBS on the beads to ensure proper bead washing.
24. Centrifuge again at 1,600 × *g* for 10 min at 4 °C. Carefully discard the supernatant.
25. Add 500 μL pre-chilled PBS to each tube. Estimate the amount of protein using the protein estimation kit and adjust the volume if needed with ice-cold PBS. Make aliquots according to your needs and suitable for use in a single experiment (we normally make 200 μL aliquots for 6 samples/experiment, based on the requirement of ~2 μg of fusion protein, which normally corresponds to 30 μL of beads per sample or well of a 6-well plate).
26. Store at −80 °C for future use (*see* **Note 21**).

3.3 RhoA/GEF Pull-Down Assay

1. Culture the eukaryotic cells in a 6-well plate in a humidified environment at 37 °C with 5% CO_2 until they reach 90% confluency, following standard cell culture procedures. The cells can be cultured in full media or starvation media before treatment, depending on the nature of the experiment.

2. Prepare the Rho kinase lysis buffer and calculate 1 mL of buffer per sample (*see* **Note 22**).
3. Prepare four sets of microcentrifuge tubes per sample. Label the microcentrifuge tubes for cell sample as lysate (upon extraction from the plate, herein designated as #1), lysate supernatant (corresponding to the cell lysate supernatant, herein designated as #2), beads (corresponding to the active RhoA or GEF protein pool, herein designated as #3), and total (corresponding to the total RhoA of GEF protein pool, herein designated as #4). Set the centrifuge to 4 °C and cool down PBS on ice.
4. Thaw the prepared beads on ice and aliquot 2 μg fusion protein on beads or 30 μL beads/sample in the pre-chilled microcentrifuge tubes (#3) on ice.
5. Perform the experiment as planned (*see* **Note 23**). You can use a potent RhoA activator (i.e. $G\alpha_{12/13}$ ligand), as a positive control to ensure that the assay is working properly.
6. At the end of the incubation period, transfer the plate on ice, quickly aspirate the medium with suction pump, briefly wash the cells with ice-cold PBS, and add 500 μL of ice-cold Rho kinase lysis buffer. From this step onward, all the steps need to be performed on ice or 4 °C, unless otherwise indicated.
7. Scrape the cells with a cell scraper, collect the lysates, and centrifuge them at 13,200 × *g* for 5 min at 4 °C (*see* **Note 24**).
8. Transfer the supernatant to the microcentrifuge tubes of #1 group.
9. Transfer 420 μL from the supernatant (#2 group) to the microcentrifuge tubes containing the beads (#3 group), seal with parafilm, and incubate for 30 min at 4 °C under continuous rotation.
10. Take 50 μL from the remaining supernatant (#2 group) to the microcentrifuge tubes containing total proteins (#4 group) and add 12.5 μL 5× sample buffer. This is total cell lysate, which will serve as control for normalization.
11. After 30 min of incubation, spin down the beads at 1000 × *g* for 5 min at 4 °C.
12. Discard the supernatant without disrupting the beads and add 500 μL ice-cold lysis buffer with inhibitors to wash the unbound proteins.
13. Repeat the centrifugation process and carefully discard the supernatant, without disturbing the beads. Add 50 μL of 2× sample buffer.
14. Heat the samples from **steps 10** and **13** for 5 min at 99 °C and then spin down at maximum speed for 5 min.

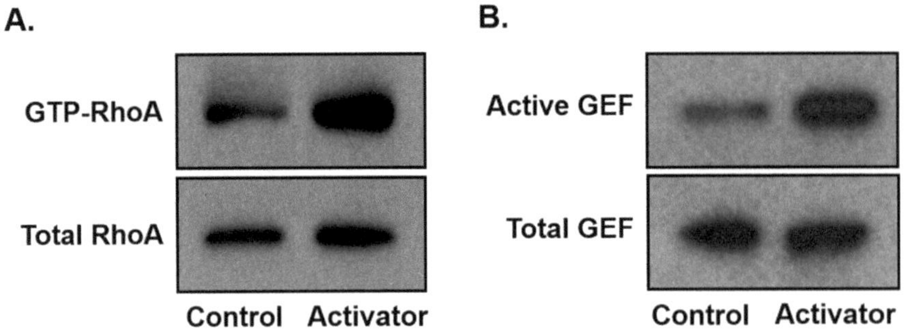

Fig. 1 (**a**) RhoA activation assay with human eukaryotic cells. Samples were prepared according to the aforementioned protocol, ran in a 15% polyacrylamide gel and the immunoreactive bands were probed with an anti-RhoA antibody. (**b**) RhoGEF activation assay with human eukaryotic cells. Samples were prepared according to the aforementioned protocol, run in a 10% polyacrylamide gel and immunoreactive bands were probed with an anti-LARG antibody

15. The samples are ready for SDS-gel electrophoresis or can be stored at −20 °C. For RhoA pull-downs, 15–18% polyacrylamide gels are used and for GEFs 8–10% polyacrylamide gels can be used.

3.4 Quantification of RhoA/GEF Activation

1. Representative Western blot images are presented in Fig. 1a (RhoA pull-down assay) and Fig. 1b (GEF activation assay).
2. The level of activation can be quantified by normalizing the bands for the bead-bound protein (active RhoA for the RhoA pull-down assay or active GEF for GEF pull-down assay) with the corresponding bands of RhoA and GEF respectively, in the total cell lysate bands.

4 Notes

1. The GST-RBD plasmid was a gift from Martin Schwartz (Addgene plasmid # 15247; http://n2t.net/addgene:15247; RRID:Addgene_15247).
2. The pGEX-4T1-RhoA G17A plasmid was a gift from Rafael Garcia-Mata (Addgene plasmid # 69357; http://n2t.net/addgene:69357; RRID:Addgene_69357).
3. Ampicillin stock with 100 mg/mL concentration in H_2O can be prepared and stored at −20 °C in small aliquots. For 10 mL stock solution, weigh 1 g of ampicillin and transfer it to 9 mL purified H_2O. Vortex to dissolve completely and adjust the volume to 10 mL. Filter sterilize by passing the solution through a 0.22-μm pore syringe filter.
4. Chloramphenicol stock with 34 mg/mL in 70% ethanol can be prepared and stored at −20 °C in small aliquots. For 10 mL

stock solution, weigh 340 mg of chloramphenicol and transfer it to 9 mL 70% ethanol. Vortex to dissolve completely and adjust the volume to 10 mL. No need to filter sterilize due to the presence of ethanol.

5. 100 mL of SOC solution can be prepared as follows:
   (a) Dissolve 2 g tryptone, 0.5 g yeast extract, 58.44 mg NaCl, and 18.638 KCl in 90 mL purified H_2O and autoclave at 121 °C for 20 min.
   (b) Dissolve 203.3 mg $MgCl_2$ and 360.31 mg glucose in 10 mL deionized H_2O and filter sterilize by passing the solution through a 0.22-μm pore syringe filter.
   (c) When the autoclaved solution (prepared in **step a**) comes to room temperature (RT), add the solution prepared in **step b**.
   (d) SOC medium can be stored at RT for further use until sterility is compromised.
6. A 10 mL IPTG stock solution of 100 mM concentration can be prepared as follows: Add 0.2383 g IPTG in 9 mL purified H_2O and vortex to dissolve. Adjust the volume to 10 mL and filter sterilize by passing the solution through a 0.22-μm pore syringe filter. Dispense the solution in 1 mL aliquots and store at −20 °C.
7. An aprotinin stock solution of 10 mg/mL concentration can be prepared as follows: Add 1 mL purified H_2O or PBS in a 10 mg vial (the quantity can be adjusted to match 100 μL of diluent to 1 mg of aprotinin) and vortex to dissolve. Dispense the solution in 100 μL aliquots and store at −20 °C.
8. A leupeptin stock solution of 10 mg/mL concentration can be prepared as follows: Add 0.5 mL purified H_2O or PBS in a 5 mg vial (the quantity can be adjusted to match 100 μL of diluent to 1 mg of Leupeptin) and vortex to dissolve. Dispense the solution in 100 μL aliquots and store at −20 °C.
9. DTT solution of 1 M concentration can be prepared as follows: Add 1.5 g of DTT in 8 mL of purified H_2O or PBS and vortex to dissolve. Adjust the volume to 10 mL, dispense the solution in 1 mL aliquots, wrap them with aluminum foil to protect from light and store at −20 °C.
10. A 10 mL PMSF stock solution of 100 mM concentration can be prepared as follows: Add 174.2 mg of PMSF in 10 mL isopropanol and vortex to dissolve. While vortexing, seal the lid of the 15-mL conical tube with parafilm to avoid evaporation. Store at −20 °C.
11. A 10 mL Na_3VO_4 stock solution of 200 mM concentration can be prepared as follows:

(a) Dissolve 368 mg of sodium orthovanadate in 9 mL purified H_2O in a conical tube and mix by vortexing.
(b) When dissolved, adjust the pH to 10. At pH 10, the solution should be yellow.
(c) Boil the solution until it becomes colorless (approximately 10 min). All crystals should dissolve.
(d) Allow the solution to come to room temperature.
(e) Measure the pH and readjust if needed.
(f) Repeat **steps** "**b**" to "**e**" until the solution remains colorless and pH stabilizes at 10.
(g) Adjust the final volume to 10 mL with purified H_2O.
(h) Dispense the solution in 500 μL aliquots and store at −20 °C.

12. For our experiments, FBS is being heat-inactivated in 56 °C for 30 min, aliquoted and stored at −20 °C.
13. 1× PBS: Add NaCl 8 g/L, KCl 0.2 g/L, Na_2HPO_4 1.44 g/L, KH_2PO_4 0.24 g/L in distilled water. Adjust pH to 7.4. Prior to use, the buffer is autoclaved at 121 °C for 20 min.
14. Ampicillin is used in the LB medium and agar plates to ensure that the bacteria containing the plasmid are selected, since both plasmids have ampicillin resistance. Chloramphenicol is used in the LB medium and agar plates to maintain the expression of the T7 polymerase plasmid that the BL21(DE3) bacteria have. This facilitates protein expression upon IPTG treatment.
15. Prior to 4 °C storage, agar plates should be sealed with parafilm and stored in an inverted position to preserve moisture.
16. The number of plates and volume can be adjusted to obtain appropriate bacterial concentrations necessary for generating single colonies. Always calculate an extra agar plate and 50 μL of nontransformed bacteria as control. This validates the efficiency of the antibiotic during the colony formation of the plasmid-containing bacteria.
17. Glycerol stocks are performed as follows: Autoclave glycerol at 121 °C for 20 min. Let it cool down to room temperature, where it will be stored. For glycerol stock formation, add 850 μL of bacterial-containing LB media and 150 μL glycerol in sterile microcentrifuge vials, briefly vortex to mix and store at −80 °C for further use.
18. Fusion proteins, especially G17A, are not stable; therefore, bead preparation should be performed at indicated temperatures without delay within the steps. Please work ahead of time preparing for future steps of the protocol, such as pre-chilling the centrifuge.

19. Some protease inhibitors lose their efficacy faster at room temperature. Thus, inhibitors should be thawed and added to buffers just before use and everything should be performed on ice.
20. The viscosity of the lysate should be lost after proper sonication. Users might need to adjust the sonication power appropriately to ensure complete lysis and have the sample immersed on ice to avoid increase of sample's temperature.
21. Freeze–thaw cycle severely affects the stability of the fusion proteins. The remaining of the beads from one experiment can only be used once afterwards.
22. *Rho Kinase Lysis Buffer Preparation*:
    (a) Prepare lysis buffer for RhoA pull-down assay by dissolving the following in 400 mL purified water:

Reagent	Final concentration
HEPES (pH 7.5)	20 mM
Triton X-100	1%
NaCl	100 mM
$MgCl_2 \cdot 6H_2O$	20 mM
EGTA	10 mM
β-Glycerol phosphate	40 mM

    (b) Adjust the pH to 7.4 and volume to 500 mL. Store at 4 °C.
    (c) Just before the assay, add the following inhibitors (working lysis buffer):

Reagent	Final concentration
PMSF[a]	1 mM
Aprotinin	10 μg/mL
Leupeptin	10 μg/mL
Na_3VO_4	1 mM
DTT	1 mM

[a]PMSF needs to thaw first so that the crystals dissolve prior to use

    (d) Throughout the assay, keep the working lysis buffer on ice.
23. During optimization of the method, it is advised to perform a time-course experiment, to identify the optimal time point for your experiment. RhoA activation can be a rapid process, starting from 30 s, but can last for 30 min or even longer [6].

24. In case the cell number varies among the groups/samples, protein estimation should be performed to equalize the concentration of protein. Attention should be paid to minimize the time spent in this process, to preserve the stability of active conformation over time.

Acknowledgments

This work was supported in part by grants from the National Institutes of Health (NCI) R15CA231339 and Texas Tech University Health Sciences Center (TTUHSC) Office of Research. The funders had no role in study design, decision to write, and preparation of the manuscript.

References

1. Ridley AJ (2015) Rho GTPase signalling in cell migration. Curr Opin Cell Biol 36:103–112. https://doi.org/10.1016/j.ceb.2015.08.005
2. Kim JG, Islam R, Cho JY et al (2018) Regulation of RhoA GTPase and various transcription factors in the RhoA pathway. J Cell Physiol 233 (9):6381–6392. https://doi.org/10.1002/jcp.26487
3. Zahra FT, Sajib MS, Ichiyama Y et al (2019) Endothelial RhoA GTPase is essential for in vitro endothelial functions but dispensable for physiological in vivo angiogenesis. Sci Rep 9(1):11666. https://doi.org/10.1038/s41598-019-48053-z
4. Desai LP, Aryal AM, Ceacareanu B et al (2004) RhoA and Rac1 are both required for efficient wound closure of airway epithelial cells. Am J Physiol Lung Cell Mol Physiol 287(6): L1134–L1144. https://doi.org/10.1152/ajplung.00022.2004
5. Jackson B, Peyrollier K, Pedersen E et al (2011) RhoA is dispensable for skin development, but crucial for contraction and directed migration of keratinocytes. Mol Biol Cell 22 (5):593–605. https://doi.org/10.1091/mbc.E09-10-0859
6. Mikelis CM, Palmby TR, Simaan M et al (2013) PDZ-RhoGEF and LARG are essential for embryonic development and provide a link between thrombin and LPA receptors and Rho activation. J Biol Chem 288 (17):12232–12243. https://doi.org/10.1074/jbc.M112.428599
7. Michaelson D, Silletti J, Murphy G et al (2001) Differential localization of Rho GTPases in live cells: regulation by hypervariable regions and RhoGDI binding. J Cell Biol 152(1):111–126. https://doi.org/10.1083/jcb.152.1.111
8. van Unen J, Reinhard NR, Yin T et al (2015) Plasma membrane restricted RhoGEF activity is sufficient for RhoA-mediated actin polymerization. Sci Rep 5:14693. https://doi.org/10.1038/srep14693
9. Garcia-Mata R, Boulter E, Burridge K (2011) The 'invisible hand': regulation of RHO GTPases by RHOGDIs. Nat Rev Mol Cell Biol 12(8):493–504. https://doi.org/10.1038/nrm3153
10. Reid T, Furuyashiki T, Ishizaki T et al (1996) Rhotekin, a new putative target for Rho bearing homology to a serine/threonine kinase, PKN, and rhophilin in the rho-binding domain. J Biol Chem 271(23):13556–13560. https://doi.org/10.1074/jbc.271.23.13556
11. Watanabe G, Saito Y, Madaule P et al (1996) Protein kinase N (PKN) and PKN-related protein rhophilin as targets of small GTPase Rho. Science 271(5249):645–648. https://doi.org/10.1126/science.271.5249.645
12. Ren XD, Kiosses WB, Schwartz MA (1999) Regulation of the small GTP-binding protein Rho by cell adhesion and the cytoskeleton. EMBO J 18(3):578–585. https://doi.org/10.1093/emboj/18.3.578
13. van Buul JD, Geerts D, Huveneers S (2014) Rho GAPs and GEFs: controling switches in endothelial cell adhesion. Cell Adhes Migr 8 (2):108–124. https://doi.org/10.4161/cam.27599
14. Rossman KL, Der CJ, Sondek J (2005) GEF means go: turning on RHO GTPases with guanine nucleotide-exchange factors. Nat Rev Mol

Cell Biol 6(2):167–180. https://doi.org/10.1038/nrm1587

15. Erickson JW, Cerione RA (2004) Structural elements, mechanism, and evolutionary convergence of Rho protein-guanine nucleotide exchange factor complexes. Biochemistry 43(4):837–842. https://doi.org/10.1021/bi036026v
16. Schmidt A, Hall A (2002) Guanine nucleotide exchange factors for Rho GTPases: turning on the switch. Genes Dev 16(13):1587–1609. https://doi.org/10.1101/gad.1003302
17. Dvorsky R, Ahmadian MR (2004) Always look on the bright site of Rho: structural implications for a conserved intermolecular interface. EMBO Rep 5(12):1130–1136. https://doi.org/10.1038/sj.embor.7400293
18. Ridley AJ, Paterson HF, Johnston CL et al (1992) The small GTP-binding protein rac regulates growth factor-induced membrane ruffling. Cell 70(3):401–410. https://doi.org/10.1016/0092-8674(92)90164-8
19. Guilluy C, Dubash AD, Garcia-Mata R (2011) Analysis of RhoA and rho GEF activity in whole cells and the cell nucleus. Nat Protoc 6(12):2050–2060. https://doi.org/10.1038/nprot.2011.411
20. Garcia-Mata R, Wennerberg K, Arthur WT et al (2006) Analysis of activated GAPs and GEFs in cell lysates. Methods Enzymol 406:425–437. https://doi.org/10.1016/S0076-6879(06)06031-9
21. Nunes KP, Rigsby CS, Webb RC (2010) RhoA/Rho-kinase and vascular diseases: what is the link? Cell Mol Life Sci 67(22):3823–3836. https://doi.org/10.1007/s00018-010-0460-1
22. Luo W, Liu CT, Yang QH et al (2014) New angle of view on the role of rho/rho kinase pathway in human diseases. Iran J Allergy Asthma Immunol 13(6):378–395
23. Fortin Ensign SP, Mathews IT, Symons MH et al (2013) Implications of Rho GTPase Signaling in Glioma cell invasion and tumor progression. Front Oncol 3:241. https://doi.org/10.3389/fonc.2013.00241
24. Rattan S, Phillips BR, Maxwell PJ (2010) RhoA/Rho-kinase: pathophysiologic and therapeutic implications in gastrointestinal smooth muscle tone and relaxation. Gastroenterology 138(1):13–18. e11-13. https://doi.org/10.1053/j.gastro.2009.11.016

Chapter 11

Wound Matrix Stiffness Imposes on Macrophage Activation

Pu Duann and Pei-Hui Lin

Abstract

The immune system depends on two major paths—the innate and the adaptive immunity. Macrophage, with its unique features as the first line of immune defense to engulf and digest invaders, serves as the key effector cells integrating those two paths. The dynamic plasticity of macrophage activation during wound repair, inflammation resolution, and tissue remodeling are emerging biomedical and bioengineering hot topics in immune function studies such as the various secretions of cytokines and chemokines and the signaling pathways with ligands and their cognate receptors. Better knowledge on how physical/mechanical and multicellular microenvironment on the modulation of macrophage functions will create innovative therapies to boost host defense mechanism and assist wound healing. In this, we describe an easy method to measure functions (gene expressions) of human and mouse macrophages in response to mechanical microenvironment changes by embedding isolated macrophages in polymerized hyaluronan gel with different wound matrix stiffness.

Key words Hyaluronan, Extracellular matrix, Matrix stiffness, Cytokines, Chemokines, Gene expression, PBMC

1 Introduction

Macrophages contribute to diverse biology including tissue development, angiogenesis, inflammation, organ injury, and repair [1, 2]. These multifunctional macrophages phenotypes are due to their dynamic adoption/polarization in responding to signals from their microenvironments [3]. Macrophages play essential roles in all phases of wound healing process, such as boosting host immune defense from microbes via promoting and resolving inflammation through inflammatory cytokines secretions, dead cells removal, supporting epithelial cell proliferation, and tissue remodeling (scar formation) after a wound occurs [4]. Macrophages respond to chemical and physical cues upon wound healing. As such, dysregulated macrophages-mediated inflammation could also lead to tissue destruction. The tissue microenvironments such as matrix property of wound stiffness have recently opened up new research avenues and therapeutic opportunities in wound healings.

Hiranmoy Das (ed.), *Wound Regeneration: Methods and Protocols*, Methods in Molecular Biology, vol. 2193, https://doi.org/10.1007/978-1-0716-0845-6_11,

Hyaluronan (HA, hyaluronic acid) is a polyanionic, nonsulfated glycosaminoglycan composed of long chain disaccharide units of GlcNAc and D-glucuronic acid polymer that repeat up to 20,000 or more times to form a linear cable with length > 20 μm and high molecular mass (>8 million Da) [5, 6]. In fact, the wide range of molecular sizes of HA dictates its different biological functions and cellular signaling [7]. HA is a major extracellular matrix (ECM) carbohydrate polymer widely distributed throughout human body and enriched in many tissues including connective, epithelial, and neural tissues, joints, and skin. In normal human body, HA constitutes 0.02% body weight (15 g in 70 kg body weight) with rapid turnover as approximately a third of the body's HA turns over daily [5]. HA is involved in many physiological and pathobiological processes such as fertilization, cell proliferation, tissue development, immune response, wound healing, atherosclerosis, angiogenesis, tissue fibrosis, and cancers [8, 9]. These versatile biological functions are related to its unique biosysnthesis process. HA is formed directly within the pericellular undersurface of the plasma membrane and extruded into extracellular space by active plasma membrane-localized hyaluronan synthase (HAS) which is transported as inactive enzyme within vesicles through the Golgi [10].

HA greatly affects tissue mechanics during wound healing as its production could escalate to 80 fold of basal level during injury and inflammation, and decline to normal level after wound remodeling [5]. It is widely accepted that cellular actions of HA are mediated by specific HA surface receptors expressed on immune cells such as CD44 [11]. Exogenous low molecular weight HA (defined as <300 kDa) was shown to enhance wound repair and increases collagen III synthesis in aged murine dermal wounds model [12]. Both matrix-bound HA content and hydrogel stiffness affect mechanical environment [13]. Moreover, high and low molecular weight HA differentially influence macrophage activation [14]. In vitro gel systems were thus designed to assess activation of macrophages infiltrating in wound with different stiffness. For example, HA polymer mixed with formulated supplemental elements to polymerize into gels with defined stiffness was used to simulate mechanical environment of stiffed wound [15].

In this protocol, we have outlined a standard protocol applying HA gel to study macrophages activation over wounds mechanical environment, such as stiffness. We demonstrated how physical changes in the stiffness of extracellular matrix (ECM) alter macrophages functions via changes in expression of inflammatory mediators. Culture of macrophages isolated from peripheral blood mononuclear cells (PBMC) on crosslinked hyalruonan hydrogels with different HA contents (stiffness) posed different inflammatory activation. Our studies confirmed that physical tension independently contributes toward macrophage activation with subsequent differential induction of gene expressions between wildtype and a mutant mouse strain. This study thus provides insight on the role of HA on macrophage function during wound healing or tissue remodeling during disease.

2 Materials

1. Peripheral blood mononuclear cells (PBMC) derived from animals (wildtype or mutant mouse strain) or human origin (*see* **Note 1**).
2. Ficoll Histopaque-1077 (colloidal silica particles with density 1.077 g/mL, Sigma 10771) for PBMC isolation. Allow to come to room temperature before use.
3. Cell counting (hematocytometer and 0.4% trypan blue).
4. HyStem®-C Hydrogel Kits (from Advanced BioMatrix) (*see* **Note 2**).
5. RNA extraction with TRIzol (Life Technologies) and RNeasy Mini Kit (Qiagen).
6. c-DNA synthesis with reverse transcription and gene expression polymerase chain reaction (PCR) arrays (various vendors).
7. Standard tissue culture media RPMI 1640, fetal bovine serum (FBS), and penicillin and streptomycin.
8. Confocal microscope LSM 780 (Zeiss).

3 Methods

3.1 PBMC and Macrophage Isolation

1. Collect whole blood, roughly 1 mL per mouse, into a heparinized tube (*see* **Note 3**).
2. In a biosafety hood, add three parts (by volume) sterile RPMI 1640 medium to dilute the whole blood in a 15-mL Falcon conical tube.
3. Pour 4 mL Ficoll Histopaque-1077 into a 15-mL conical tube for each blood sample isolation.
4. Carefully layer 4 mL diluted blood on top of each Ficoll layer, taking care not to mix blood and Ficoll (lay Falcon conical tube almost horizontal and slowly layer the diluted blood over Ficoll).
5. Perform density gradient centrifugation with a horizontal rotor. Centrifuge at 400 × *g* for 30 min at room temperature (15–20 °C) with deceleration and acceleration both set at 1, brake off.
6. Following centrifugation, carefully remove the top plasma layer using a Pasteur pipette and discard.

7. Gently aspirate the middle PBMC layer using a Pasteur pipette and add to a new Falcon tube.
8. Wash PBMCs by adding 10 mL sterile RPMI 1640 and centrifuging at 300 × *g* for 10 min at RT.
9. Carefully remove supernatant with a pipette to avoid disturbing the loose pellet.
10. Repeat washing step once.
11. Resuspend cell pellet in 10 mL of complete culture media (RPMI 1640–10% FBS).
12. Enrich macrophages from PBMC preparation by adherence method (**steps 13** and **14**) (*see* **Note 4**).
13. Resuspend the PBMCs (roughly10–15 × 10^6 cells) and seed cells into T-75 flask.
14. Allow to adhere in a 5% CO_2 incubator at 37 °C for 2 h in 10 mL of complete culture media (RPMI 1640–10% FBS).
15. Remove nonadherent cells and carefully wash the adherent cells twice with RPMI 1640–10% FBS.
16. Verify cell viability and numbers (by 0.4% trypan blue exclusion method) under microscopy with hemocytometer or alternative cell counter.

3.2 Preparation of Hyaluronan Hydrogel Components with Optimal Stiffness

The desired stiffness of hydrogel is achieved by crosslinking various HA contents. In the current study, we focused on two HA stiffness groups, low (equivalent to Young's modulus 70 Pa) and high (110 Pa) [15]. A blank control was included with macrophages cultured in growth medium only (zero stiffness). Table 1 serves as an example of 24-well tissue culture format. Meanwhile, HA hydrogel preparation should be coordinated with "macrophage plating" (Subheading 3.3) to maximize cell viability and wellness in casted gel.

1. Prepare the working solutions for each of the three components from HyStem®-C Hydrogel Kits (*see* **Note 5**).
   Formulate HA gels accordingly (*see* Table 1).

   Low stiffness group: 1% Glycosil: 1% Gelin-S: 1% Extralink in volume ratio of 1:1:8.

   High stiffness group: 2% Glycosil: 2% Gelin-S: 2% Extralink in volume ratio of 1:1:8.

2. Reconstitute and prepare 1% and 2% Thiolated collagen (Gelin-S) solution in sterile ddH_2O following manufacturer's protocol. Reconstitute each vacuumed vial with lyophilized Gelin-S powder with 0.5- or 1- volume/mL ddH_2O which would give rise to either 2% or 1%, respectively.
3. Reconstitute and prepare 1% and 2% Sodium HA solution (Glycosil) in sterile ddH_2O by reconstituting each vacuumed

Table 1
Formulations of blank, low, and high stiffness HA gel

Macrophage in RPMI 1640–10% FBS	100 μL	100 μL	100 μL
Glycosil (100-kDa hyaluronate)	0	50 μL (1%)	50 μL (2%)
Gelin-S (thiolated collagen)	0	50 μL (1%)	50 μL (2%)
Extralink ([a]PEGDA-crosslinker)	0	400 μL (1%)	400 μL (2%)
RPMI 1640-10 FBS	750 μL	250 μL	250 μL
Stiffness[b] (pa)	Blank (0 pa)	Low (70 pa)	High (110 pa)

[a]PEGDA-Polyethylene Glycol Diacrylate Acrylate
[b]Stiffness is a mechanical property that a solid material preserves its shape resisting to stress applied on it. It is measured according to Young's modulus, derived as the reciprocal of the stress, presented in unit of Pa. In SI base units, $Pa = kg\ m^{-1}\ s^{-2}$

vial with lyophilized Glycosil powder with 0.5- or 1- volume/mL ddH_2O which would give rise to either 2% or 1%, respectively.

4. Prepare the two "Glycosil: Gelin-S mixtures" (Table 1) and set aside. First, Mix 1% Gelin-S (from Subheading 3.2, **step 2**) with 1% glycosil (from Subheading 3.2, **step 3**) at 1:1 ratio by volume. Second, mix 2% Gelin-S with 2% glycosil at 1:1 ratio by volume.
5. Reconstitute and prepare 1% and 2% polyethylene glycol diacrylate (PEGDA) and set aside. Reconstitute PEGDA powder in vial with ddH_2O to generate 1% or 2% solutions. It will require four parts (by volume) of PEGDA for each one part (volume) of "Glycosil:Gelin-S mixture."

3.3 Macrophage Plating and HA Hydrogel Polymerization

1. Prepare macrophage at a cell density of $1–5 \times 10^5$ cells per well (24-well culture plate) (*see* **Note 6**).
2. The hydrogel which will crosslink its constituents must be in a complete fluidity form when macrophages are mixed in.
3. Gently mix macrophages with the "Glycosil: Gelin-S mixture" (from Subheading 3.2, **step 4**) according to Table 1; ensure well mixing. This final mixture, in volume, contains macrophages, Glycosil, and Gelin-S at a ratio of 2:1:1.
4. Initiate polymerization by adding crosslinker PEGDA (from Subheading 3.2, **step 5**) into above mixture (from Subheading 3.3, **step 3**) with gentle pipetting (*see* **Note 7**).
5. Gel formation with embedded macrophages will proceed in sterile condition in cell culture incubator at 37 °C, 5% carbon dioxide with saturated humidity for 30 min.

A

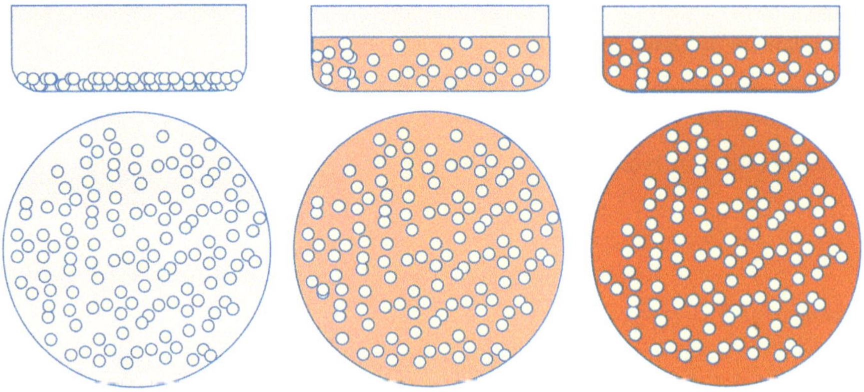

Macrophage	100 μl (1 X 10 e5)	100 μl (1 X 10 e5)	100 μl (1X 10 e5)
Glycosil	0	50 μl (1%)	50 μl (2%)
Gelin-S	0	50 μl (1%)	50 μl (2%)
Extralink	0	400 μl (1%)	400 μl (2%)
RPMI1640	750 μl	250 μl	250 μl
Stiffness (Pa)	blank control (0 Pa)	low stiffness (70 Pa)	high stiffness (110 Pa)

B C

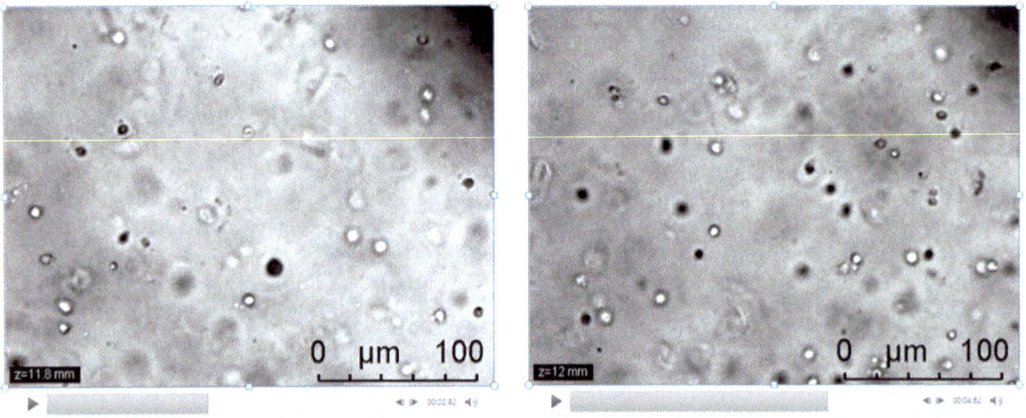

Fig. 1 3D z-stack images of plated macrophages embedded in hyaluronan gel. HA gel could be formulated with various stiffness. (**a**) Schematic diagram depicts three various formulations to prepare three HA hydrogels (blank, low, or high stiffness), respectively. Diagrams shown side (top panels) and top (middle panels) views of cultured macrophages with different HA formulations (bottom panels). (**b** and **c**) Two representative z-stack microscopic images captured from a high-stiffness gel at different z-levels (11.8-mm and 12-mm for **b** and **c**, respectively). Macrophages were shown evenly embedded in HA gels with nearly identical cell density at different z-levels. Scale bars, 100 μm

6. Replenish wells with 250 μL fresh culture medium (RPMI 1640, FBS10%) and culture for 72 h (*see* **Note 8** and Fig. 1a).
7. Image macrophages in HA hydrogels with confocal microscope (Fig. 1b, c).

3.4 Gene Expression Analysis with RT-PCR Arrays

1. Remove top layer of culture media.
2. Retrieve gel. Cells retrieving and lysis are done by applying 120 μL of TRIzol to gently crush the HA hydrogel by applying gentle force and pipetting (*see* **Note 9**).
3. Proceed with total RNA extraction and purification with TRIzol reagent and RNeasy Mini Kit according to manufacturer's protocols.
4. Measure RNA concentration using a NanoDrop spectrophotometer and use 500–1000 ng RNA to synthesize cDNA using high-capacity RNA to cDNA kit according to manufacturer's protocols. The cDNA can be stored at −20 °C until PCR reaction (*see* **Note 10**).
5. Dilute cDNA 10× with PCR-grade water.
6. Set up PCR reaction with RT^2 PCR array analysis according to manufacturer's protocols (*see* **Note 11**).
7. Alternatively, PCR could be performed with designed target primers (*see* **Note 12**).
8. Run RT-PCR reaction with target-specific PCR program.
9. Calculate expression of IL-1α, IL-1β, IL-5, MMP2 and MMP3 using the $\Delta\Delta C_T$ method. Use average expression of the three housekeeping genes (GAPDH, actin, and HSP90) to calculate the relative expression (Fig. 2).

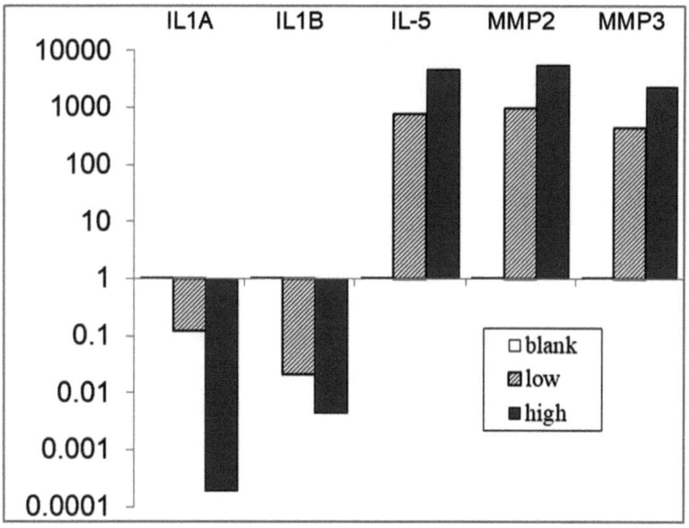

Fig. 2 Stiffness of HA hydrogel affects gene expressions of isolated macrophages derived from wildtype and mutant mice. Differential gene expressions (RT^2 PCR array) after 72-h culture under different HA gel stiffness were detected from macrophages of wildtype and mutant strain, e.g., wound-prone (IL10−/−) mice. Changes of gene expression (relative to wildtype) were indicated by folds ($2^{-\Delta\Delta Ct}$, in logarithm). Compared to wildtype macrophages, mutant macrophages displayed aberrant cytokines and MMP gene expressions in response to different microenvironmental HA stiffness

4 Notes

1. The PBMC could be obtained from either mouse or human origins. PBMCs contain enriched leukocytes and monocytes. This method could be applied to isolate macrophages from wildtype or mutant murine strains with immune dysfunctions [16]. Alternatively, heparinized human whole blood (65 mL) could be obtained from normal healthy donors (sourced from blood bank) or specific patient targets [17].
2. HyStem®-C Hydrogel Kits contains three biocompatible components: thiol-modified hyaluronan (Glycosil) (100 kDa), thiol-modified denatured collagen (Gelin-S®), and thiol-reactive crosslinker, PEGDA (Extralink). Each is packaged as sterile lyophilized powder.
3. We demonstrated mouse whole blood PBMC isolation here. For human PBMC isolation, scale up reagents accordingly. In general, routinely 1–2 $\times$ 10^6 PBMC will be obtained per mL blood. Assuming 10% monocyte content, 2.5–3 $\times$ 10^7 monocytes is expected to be isolated from 250 mL whole blood. Compared with normal subjects, abnormal macrophages behaviors might be detected from patients with remarkable keloid scar condition.
4. A typical yield would be achieved for monocyte (assuming 10–25% in content) by seeding 1.0 $\times$ 10^6 PBMC per cm^2 of adhesive area. An alternative to isolate monocytes is affinity column purification where monocyte/macrophage will be retained in affinity column, while the rest of the cells elute off [17]. Follow manufacturer's protocol accordingly.
5. We followed gel preparation method according to *Nagy* et al. ([15]) to achieve three different (blank control, low and high) gel stiffness by mixing sodium hyaluronate (HA) in thiolated collagen (Gelin-S) and crosslinker PEGDA.
6. One mL of mouse peripheral blood would yield 1–7 $\times$ 10^6 PBMC and 25% of them are macrophage/monocytes. A practical preparation is to pool total recovery macrophages from plastic adherence of mouse PBMC, make a master mixing, and reconstitute to 0.5 mL with culture media (RPMI 1640–10% FBS). Divide those concentrated cells to 6–8 aliquots. Each aliquot, containing approximately 1–5 $\times$ 10^5 monocyte/macrophages, is subjected to each experimental group analysis.
7. Gently pipette-mixing the cells in all three hydrogel components would ensure even distribution of cells embedded in gel from bottom to top. This would ensure that macrophages respond to only calculated gel stiffness in all three experimental groups. This step is critical and might require several practices to achieve perfection.

8. Macrophages are cultured in basal medium (RPMI 1640 with 10% FBS, glutamine, and antibiotics penicillin/streptomycin) to ensure no further activation from growth factors such as M-CSF (myeloid-colony stimulating factor (M-CSF) and only respond to physical environmental cue.
9. TRIzol contains corrosive phenol and should be handled with caution in a chemical hood or a biosafety hood.
10. Use 500–1000 ng total RNA to synthesize cDNA in a final 20-μL reaction mixture.
11. Several target RT^{2}PCR arrays are commercially available, such as inflammation, fibrosis cytokines, and chemokines arrays.
12. Alternatively, PCR gene expression analysis could be performed with self-designed target primers. Mix 5 μL (2×) Fast Sybr Green Master Mix, 0.5 μL forward and reverse primers (from 10 μM stock, the final primer concentration is 500 nM), and 4 μL diluted cDNA (equivalent to original 50–100 ng original RNA) for each 384-well final RT-PCR reaction. Final volume is 10 μL per well.

Acknowledgments

This work was supported by an Ohio State University intramural Lockwood Research grant (to P.H.L).

References

1. Hamidzadeh K, Christensen SM, Dalby E et al (2017) Macrophages and the recovery from acute and chronic inflammation. Annu Rev Physiol 79:567–592. https://doi.org/10.1146/annurev-physiol-022516-034348
2. Vannella KM, Wynn TA (2017) Mechanisms of organ injury and repair by macrophages. Annu Rev Physiol 79:593–617. https://doi.org/10.1146/annurev-physiol-022516-034356
3. Murray PJ (2018) Immune regulation by monocytes. Semin Immunol 35:12–18. https://doi.org/10.1016/j.smim.2017.12.005
4. Lin PH, Sermersheim M, Li H et al (2017) Zinc in wound healing modulation. Nutrients 10(1). https://doi.org/10.3390/nu10010016
5. Fraser JR, Laurent TC, Laurent UB (1997) Hyaluronan: its nature, distribution, functions and turnover. J Intern Med 242(1):27–33. https://doi.org/10.1046/j.1365-2796.1997.00170.x
6. Hascall VC (2019) The journey of hyaluronan research in the journal of biological chemistry. J Biol Chem 294(5):1690–1696. https://doi.org/10.1074/jbc.TM118.005836
7. Tavianatou AG, Caon I, Franchi M et al (2019) Hyaluronan: molecular size-dependent signaling and biological functions in inflammation and cancer. FEBS J 286(15):2883–2908. https://doi.org/10.1111/febs.14777
8. Jiang D, Liang J, Noble PW (2011) Hyaluronan as an immune regulator in human diseases. Physiol Rev 91(1):221–264. https://doi.org/10.1152/physrev.00052.2009
9. Spinelli FM, Vitale DL, Demarchi G et al (2015) The immunological effect of hyaluronan in tumor angiogenesis. Clin Transl Immunology 4(12):e52. https://doi.org/10.1038/cti.2015.35
10. Weigel PH, Hascall VC, Tammi M (1997) Hyaluronan synthases. J Biol Chem 272 (22):13997–14000. https://doi.org/10.1074/jbc.272.22.13997

11. Johnson P, Ruffell B (2009) CD44 and its role in inflammation and inflammatory diseases. Inflamm Allergy Drug Targets 8(3):208–220
12. Damodarasamy M, Johnson RS, Bentov I et al (2014) Hyaluronan enhances wound repair and increases collagen III in aged dermal wounds. Wound Repair Regen 22 (4):521–526. https://doi.org/10.1111/wrr.12192
13. Chen JE, Pedron S, Harley BAC (2017) The combined influence of hydrogel stiffness and matrix-bound hyaluronic acid content on Glioblastoma invasion. Macromol Biosci 17(8). https://doi.org/10.1002/mabi.201700018
14. Rayahin JE, Buhrman JS, Zhang Y et al (2015) High and low molecular weight hyaluronic acid differentially influence macrophage activation. ACS Biomater Sci Eng 1(7):481–493. https://doi.org/10.1021/acsbiomaterials.5b00181
15. Nagy N, de la Zerda A, Kaber G et al (2018) Hyaluronan content governs tissue stiffness in pancreatic islet inflammation. J Biol Chem 293 (2):567–578. https://doi.org/10.1074/jbc.RA117.000148
16. Zhang X, Goncalves R, Mosser DM (2008) The isolation and characterization of murine macrophages. Curr Protoc Immunol. Chapter 14:Unit 14.11. https://doi.org/10.1002/0471142735.im1401s83
17. Kelly A, Grabiec AM, Travis MA (2018) Culture of human monocyte-derived macrophages. Methods Mol Biol 1784:1–11. https://doi.org/10.1007/978-1-4939-7837-3_1

Chapter 12

Generation of Acute Hind Limb Ischemia in NOD/SCID Mice and Treatment with Nanofiber-Expanded CD34⁺ Hematopoietic Stem Cells

Derek Barthels and Hiranmoy Das

Abstract

Critical limb ischemia (CLI) is primarily associated with a high risk of major amputation, cardiovascular events, and death. The current therapy involves direct endovascular intervention and is associated with long-term recurrence. However, patients with significant comorbidities are not eligible for this therapy. Hind limb ischemia model via femoral artery excision has commonly been used to determine therapeutic potential and for investigating cellular and molecular mechanisms. This protocol describes the ischemic model development in NOD/SCID mice and the use of human umbilical cord blood-derived and nanofiber scaffold-expanded $CD34^+$ stem cells to investigate the efficacy of regenerative therapy.

Key words Peripheral artery disease, Critical limb ischemia, Femoral artery excision, $CD34^+$ stem cells

1 Introduction

Peripheral artery disease (PAD) is the categorization for atherosclerotic occlusive conditions that affect large or medium-sized arteries outside of cardiac or cerebral circulation [1]. PAD is a major contributor to vascular morbidity and is often associated with lower limb ischemia in conditions such as diabetes mellitus [2, 3]. If left unchecked, PAD can advance to critical limb ischemia (CLI), which consists of ischemic rest pain, nonhealing ulcers, and tissue loss. These conditions lead to secondary complications, including amputations, increased odds of cardiovascular events, and a 50% mortality rate within 5 years [4, 5]. The outlook for diabetic patients is even more severe. As of 2015, between 40% and 50% of diabetic patients undergo amputations and 20% to 25% die within 1 year of a CLI diagnosis [6]. While revascularization of the tissue in question is undoubtedly the final objective for CLI treatments, there is some ambiguity as to which strategy is best to

Hiranmoy Das (ed.), *Wound Regeneration: Methods and Protocols*, Methods in Molecular Biology, vol. 2193,
https://doi.org/10.1007/978-1-0716-0845-6_12, © Springer Science+Business Media, LLC, part of Springer Nature 2021

achieve this goal, as clinical evidence of efficacy remains elusive [4]. The current first-line treatments for CLI involve direct endovascular intervention. However, these invasive treatments may not be viable for patients with significant comorbidities, and long-term recurrence rates are high after these treatments are administered alone [5]. Because of the severity accompanied by CLI and its devastating effect on patients' quality of life, the development of suitable animal models to recapitulate the disease state is sorely needed so that new, more promising treatments can be developed.

The application of stem cells has been a large focus in developing regenerative therapies for several illnesses. Human adult stem cells have been used in a large number of studies investigating stem cell therapies in various disease states, including chronic wound healing [7] and osteoporosis [8, 9]. Nanofiber-expanded $CD34^+$ stem cells, in particular, have shown promise in revascularization of ischemic tissue in a model of CLI in nonobese diabetic (NOD)/severe combined immunodeficient (SCID) mice [9]. This study highlights the importance of the development of animal models in multiple genetic strains of mice, as allogeneic transplantation of hematopoietic stem cells must consider various factors in the immune system. When performed between humans, transplantation requires processes such as human leukocyte antigen (HLA) matching, and while matching possible donors in humans is challenging, trans-species transplantation of hematopoietic stem cells presents vastly more immune obstacles to be overcome [10]. The development of models in immunocompromised animals is one way to navigate the obstacles that prohibit stem cell transplantation across species.

Of utmost importance in these models is the consistency and reproducibility of the results they generate. The excision of the femoral artery has been a standard method to model CLI in mice [3]. The use of this model has been instrumental in elucidating possible factors that influence physiological responses to limb ischemia, including age [11], genetic variables [12], and comorbidities such as Type 1 diabetes [13]. However, the implementation of the model in NOD/SCID mice allows for vastly more possibilities to assess efficacy and safety in preclinical conditions that are otherwise not viable in immunocompetent animals. With the future of CLI treatments still in serious doubt as to how best move forward, preclinical animal research gives a reproducible, controlled environment to explore these stem cell treatment options and factors that may influence recovery in human patients.

2 Materials

2.1 Hair Removal in Preparation for Surgery

1. NOD/SCID mice.
2. Anesthesia chamber.
3. Heating pad with nosecone for isoflurane.
4. Ophthalmic lubricant.
5. Electric trimmer.
6. Hair removal cream.

2.2 Ligation and Excision of the Femoral Artery

1. Heating pad.
2. Surgical scissors.
3. Surgical forceps.
4. Surgical tape.
5. Tissue retractors.
6. Cotton swabs and gauze.
7. Dissection microscope.
8. Ketamine HCl (100 mg/kg)/xylazine (10 mg/kg) cocktail.
9. Temperature probe.
10. Povidone–iodine and 70% alcohol solutions.
11. 7—0 silk sutures
12. 5—0 silk sutures.

2.3 Isolation, Expansion, and Transplantation of Human Umbilical Cord Blood-Derived Hematopoietic Stem Cells

1. AutoMACS cell sorter.
2. Flow cytometer.
3. Nanofiber scaffolds [14].
4. StemSpan™ serum-free expansion media (SFEM) [9].
5. Nontrypsin dissociating agent.
6. Hanks' buffer.

2.4 Analysis of Functional Recovery Using Laser Doppler Perfusion Imaging

1. Anesthesia chamber.
2. Heating pad with nosecone for isoflurane.
3. Laser Doppler perfusion image analyzer.

2.5 Determination of Capillary Density in Ischemic Tissue

1. Alkaline phosphate kit.
2. Microscope.

3 Methods

3.1 Removal of Hair from the Site of Operation

All surgical preparations and procedures should be performed on a covered heating pad set to 37 °C to ensure the body temperatures of the mice remain constant while under anesthesia (*see* **Notes 1–6**).

1. Place the mouse in the anesthesia chamber with 2–5% isoflurane flow. Ensure the mouse is anesthetized by gently rocking the anesthesia chamber. If there is no response, a toe pinch can be performed to confirm anesthesia further. There should be no reaction from the mouse.
2. Move the mouse from the anesthesia chamber to the heating pad. Place the dorsal side of the mouse down and set its nose in the isoflurane port connected to the heating pad.
3. While gently extending each hind leg of the mouse, shave the hair from each limb while being careful not to cut the skin.
4. Warm some hair removal cream and apply it to each limb. Gently wipe cream and hair away after 1 min.
5. Move the mouse back to the cage and monitor recovery from anesthesia.

3.2 Ligation and Transection of the Femoral Artery (Fig. 1)

1. Induce anesthesia using an intraperitoneal 0.1 mL/20 g injection of the ketamine/xylazine cocktail. Perform a toe pinch to ensure that anesthesia is complete. The mouse should not react.
2. Place the mouse on the covered heating pad, dorsal side down. Extend the mouse's limbs and tape them to the pad. Insert the temperature probe in the rectum of the mouse and tape it to the heating pad to ensure it does not move during surgery.
3. Disinfect surgical site using alternating wipes of a povidone–iodine solution and a 70% ethanol solution.
4. Using forceps to lift the skin, make an incision lengthwise down the center of the mouse's thigh, between the mouse's body and knee, with sharp surgical scissors.
5. Use forceps to widen the incision. Carefully insert surgical retractors to hold open the surgical site.
6. The femoral artery and the femoral vein will both be visible under a membrane once the fat tissue under the skin is pulled aside. Carefully insert forceps under the membrane and let the forceps open to pull it apart.
7. The femoral artery runs parallel and ventral to the femoral vein. Gently place forceps between the two vessels and guide them apart to isolate the artery. Repeat this process to isolate the artery from the nerve that runs along with it.

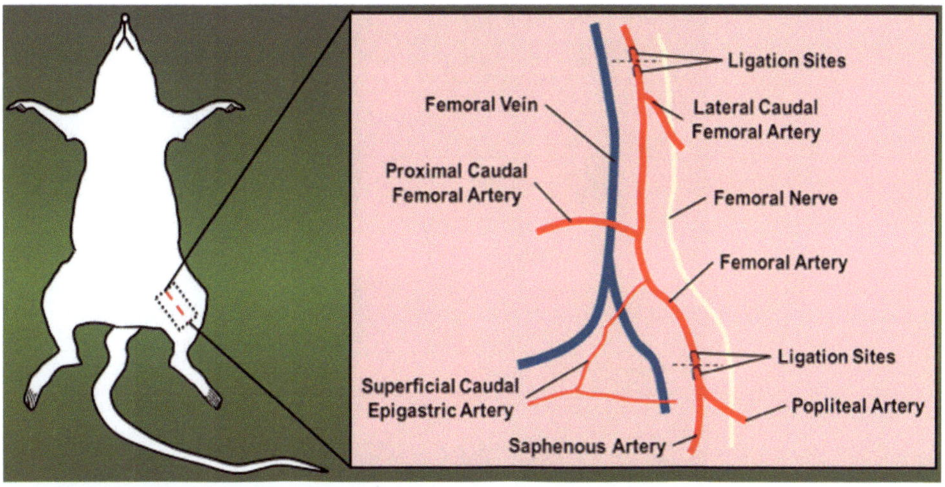

Fig. 1 Vascular features in the murine hind limb. A pair of ligations will be tied just proximal to the lateral caudal femoral artery bifurcation, while the other set will be placed just proximal to the bifurcation of the saphenous and popliteal arteries. The artery will be transected between each set of these ligations as shown

8. Run 7—0 sutures under the artery. Tie two sutures at the proximal end of the artery and two sutures at the distal end of the artery. The distal pair of sutures should be just proximal to the popliteal and saphenous arteries.
9. Transect the artery between each pair of sutures and close the wound using 5—0 sutures.
10. Use perfusion radar Doppler imaging to confirm the ischemia of the limb (*see* **Note 7**). After confirmation, move the mouse to a recovery chamber and monitor its recovery from anesthesia (Fig. 1).

3.3 Isolation, Expansion, and Transplantation of CD34+ Stem Cells

1. Using Ficoll-Paque centrifugation, isolate mononuclear cells from umbilical cord blood.
2. Use an AutoMACS cell sorter to isolate $CD133^+$ cells from the resulting mononuclear cell population.
3. Perform flow cytometry to confirm the presence of a homogenous, $CD133^+$ cell population.
4. To expand the hematopoietic stem cells, glue nanofiber scaffolds to the bottom of wells in a 24-well tissue culture dish (*see* **Note 8**). Plate 800 $CD133^+$ cells resuspended in SFEM in each well and culture for 10 days at 37 °C and 5% CO_2 without changing media (*see* **Note 9**).
5. At the end of the 10-day culture period, wash each well with a nontrypsin dissociating agent once, followed by two rinses using Hanks' buffer containing 2% fetal bovine serum (FBS). There should be a minimum of 5 min between each rinse.

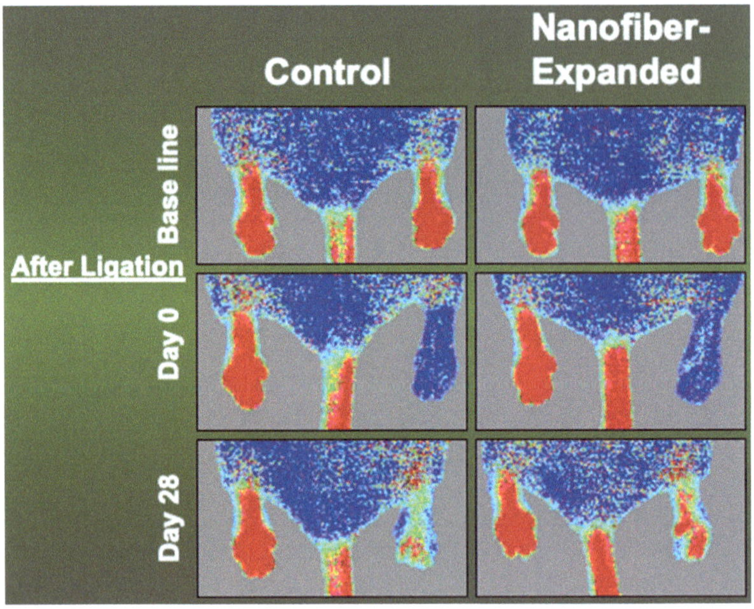

Fig. 2 Laser Doppler perfusion imaging before and after femoral artery ligation and treatment with nanofiber-expanded umbilical cord blood-derived stem cells. Blood flow to the hind limbs was visualized by Laser Doppler perfusion imaging before and after the excision of the femoral artery. Revascularization was examined weekly in both control and treatment groups by perfusion imaging until the endpoint of the study

6. Collect the resulting cell suspensions, centrifuge at 500 × *g* for 10 min, and resuspend in media at a density of 5×10^5 cells/300 μL.
7. Inject 5×10^5 cells in a volume of 300 μL via intraventricular injection.
8. Move the mouse to a recovery chamber and monitor its recovery from the procedure before moving it back to its cage.

3.4 Analysis of Functional Recovery Using Laser Doppler Perfusion Imaging

Anesthetize the mouse as was done in Subheading 3.1. Collection of Laser Doppler images of the affected limb should be done before and after the excision of the femoral artery. Images should also be collected 7, 14, 21, and 28 days after stem cell transplantation (Fig. 2).

3.5 Determination of Capillary Density in Ischemic Tissue

1. Dissect and snap freeze tissue from the ischemic site of the hind limb.
2. Stain tissue sections using an alkaline phosphatase kit.
3. An investigator blinded to the study selects four sections within the necrosis border zone and counts stained capillaries within 12 randomly selected high-power fields. Final counts can be expressed as capillaries per high power field.

4 Notes

1. Researchers performing this procedure must be trained to work in a BSL-2 lab.
2. Experiments listed are performed using 7-week-old, NOD/SCID mice. All animal procedures must be performed in accordance with an approved Institution of Animal Care and Use Committee (IACUC) protocol.
3. The dosage of the ketamine/xylazine mixture is extremely important. If too little of the cocktail is injected, the mouse will not be properly anesthetized for the operation. If too much of the cocktail is injected, it could cause respiratory depression, resulting in the death of the mouse. Similar issues may occur if the ketamine/xylazine cocktail is not formulated correctly.
4. While not required, the use of a temperature probe is highly recommended to better monitor the body temperature and well-being of anesthetized animals.
5. To help visualize the necessary anatomical landmarks during the surgery, the procedure should be performed under a dissection microscope.
6. After surgery, mice should be monitored daily for signs of excessive pain or stress. Mice showing pain or stress deemed excessive by the IACUC protocol's endpoints should be humanely euthanized by way of isoflurane and cervical dislocation.
7. Laser Doppler imaging should be performed immediately after surgery to confirm limb ischemia.
8. For SFEM components and supplements, *see* [9].
9. For the synthesis of nanofiber scaffolds, *see* [14].

Acknowledgments

This work was supported in part by National Institutes of Health grants, R01AR068279 (NIAMS), STTR 1R41EY024217 (NEI), and STTR 1R41AG057242 (NIA).

References

1. Krishna SM (2019) Vitamin D as a protector of arterial health: potential role in peripheral arterial disease formation. Int J Mol Sci 20(19). https://doi.org/10.3390/ijms20194907
2. Edmonds M (2019) Vascular disease in the lower limb in type 1 diabetes. Cardiovasc Endocrinol Metab 8(1):39–46. https://doi.org/10.1097/XCE.0000000000000168
3. Padgett ME, McCord TJ, McClung JM et al (2016) Methods for acute and subacute murine Hindlimb ischemia. J Vis Exp 112. https://doi.org/10.3791/54166

4. Farber A, Eberhardt RT (2016) The current state of critical limb ischemia: a systematic review. JAMA Surg 151(11):1070–1077. https://doi.org/10.1001/jamasurg.2016.2018
5. Gupta A, Lee MS, Gupta K et al (2019) A review of antithrombotic treatment in critical limb ischemia after endovascular intervention. Cardiol Ther 8(2):193–209. https://doi.org/10.1007/s40119-019-00153-7
6. Elsayed S, Clavijo LC (2015) Critical limb ischemia. Cardiol Clin 33(1):37–47. https://doi.org/10.1016/j.ccl.2014.09.008
7. Kanji S, Das H (2017) Advances of stem cell therapeutics in cutaneous wound healing and regeneration. Mediat Inflamm 2017:5217967. https://doi.org/10.1155/2017/5217967
8. Rolph DN, Deb M, Kanji S et al (2018) Ferutinin directs dental pulp-derived stem cells towards the osteogenic lineage by epigenetically regulating canonical Wnt signaling. Biochim Biophys Acta Mol basis Dis 1866 (4):165314. https://doi.org/10.1016/j.bbadis.2018.10.032
9. Das H, Abdulhameed N, Joseph M et al (2009) Ex vivo nanofiber expansion and genetic modification of human cord blood-derived progenitor/stem cells enhances vasculogenesis. Cell Transplant 18(3):305–318. https://doi.org/10.3727/096368909788534870
10. Tiercy JM (2016) How to select the best available related or unrelated donor of hematopoietic stem cells? Haematologica 101 (6):680–687. https://doi.org/10.3324/haematol.2015.141119
11. Okadome J, Matsumoto T, Yoshiya K et al (2018) BubR1 insufficiency impairs angiogenesis in aging and in experimental critical limb ischemic mice. J Vasc Surg 68(2):576–586.e571. https://doi.org/10.1016/j.jvs.2017.07.119
12. McClung JM, McCord TJ, Keum S et al (2012) Skeletal muscle-specific genetic determinants contribute to the differential strain-dependent effects of hindlimb ischemia in mice. Am J Pathol 180(5):2156–2169. https://doi.org/10.1016/j.ajpath.2012.01.032
13. Peravali R, Gunnels L, Alleboina S et al (2019) Type 1 diabetes alters ischemia-induced gene expression. J Clin Transl Endocrinol 15:19–24. https://doi.org/10.1016/j.jcte.2018.11.003
14. Chua KN, Chai C, Lee PC et al (2006) Surface-aminated electrospun nanofibers enhance adhesion and expansion of human umbilical cord blood hematopoietic stem/progenitor cells. Biomaterials 27(36):6043–6051. https://doi.org/10.1016/j.biomaterials.2006.06.017

Chapter 13

Electrospun Aligned Coaxial Nanofibrous Scaffold for Cardiac Repair

Divya Sridharan, Arunkumar Palaniappan, Britani N. Blackstone, Heather M. Powell, and Mahmood Khan

Abstract

Cardiovascular diseases (CVDs) are one of the leading causes of mortality worldwide and a number one killer in the USA. Cell-based approaches to treat CVDs have only shown modest improvement due to poor survival, retention, and engraftment of the transplanted cells in the ischemic myocardium. Recently, tissue engineering and the use of 3D scaffolds for culturing and delivering stem cells for ischemic heart disease are gaining rapid potential. Here, we describe a protocol for the fabrication of aligned coaxial nanofibrous scaffold comprising of a polycaprolactone (PCL) core and gelatin shell. Furthermore, we describe a detailed protocol for the efficient seeding and maintenance of human-induced pluripotent stem cell-derived cardiomyocytes (hiPSC-CMs) on these nanofibrous scaffolds, which could have a potential application in the generation of functional "cardiac patch" for myocardial repair applications as well as an in vitro 3D cardiac tissue model to evaluate the efficacy of cardiovascular drugs and cardiac toxicities.

Key words Electrospinning, Nanofibrous scaffolds, Cardiomyocytes, Cardiac repair

1 Introduction

Cell-based approaches have become an integral part in the field of regenerative medicine, especially for organs with limited or no regenerative potential, for example, the heart. Although a number of different cell types, including various stem/progenitor cells as well as cardiomyocytes, have been studied, only a modest improvement in cardiac function has been reported following transplantation into the ischemic myocardium. A vast majority of the studies have attributed this to poor retention and low survival of the cells at the site of infarction [1]. To overcome these limitations, tissue-engineering techniques have been employed in combination with cells to improve the ease of cell delivery as well as retention [2]. Among these techniques, use of 3D bioengineered scaffolds in combination with different cell types has become increasingly popular. Such scaffolds have been successfully used in various

Hiranmoy Das (ed.), *Wound Regeneration: Methods and Protocols*, Methods in Molecular Biology, vol. 2193,
https://doi.org/10.1007/978-1-0716-0845-6_13,

preclinical [3] as well as clinical studies [4]. However, several factors have been shown to influence the successful use of such bioengineered scaffolds for cardiac applications, namely, the nature of biomaterial used, mechanical properties of the scaffolds, thickness and dimensions, and alignment [2, 5].

Studies have reported the use of natural biomaterials such as collagen, gelatin, and fibrin in scaffolds for culturing cardiomyocytes [6]. While these materials do not alter cell viability or function, they have been shown to exhibit poor strength and stiffness [7]. On the other hand, synthetic polymers like polycaprolactone (PCL) and poly (lactic-*co*-glycolic acid) (PLGA) have been used to fabricate scaffolds with better mechanical properties, but the adhesion and survival of cells on these patches were low [8, 9]. Coaxial (CoA) nanofiber systems, with a synthetic polymer (PCL or PLA) in the core and a more adhesive natural material (gelatin) as the shell, exhibited both mechanical strength, high cell adhesion and viability, and lower in vitro production of inflammatory cytokines when compared to purely synthetic nanofibrous scaffold [8, 10]. For tissue-engineering applications, among various available techniques, electrospinning has been extensively used for fabricating nanofibrous scaffolds [2]. In this chapter, we present a method for the fabrication of an aligned polycaprolactone (PCL)–gelatin co-axial nanofibrous scaffold for the development of cardiac patch, which could be implanted along with hiPSC-CMs on the epicardial heart surface for potential cardiac tissue regeneration.

2 Materials

Prepare all reagents using sterile ultrapure water and culture-grade reagents. Prepare and store all reagents at room temperature (until mentioned otherwise). Prepare all organic solvents in a chemical hood and use the appropriate personal protective equipment.

2.1 Electrospinning

1. 12 wt./vol.% Gelatin solution: Dissolve 1.2 g of gelatin from bovine skin in 10 ml of 1,1,1,3,3,3-hexafluoro-2-propanol (HFIP) (*see* **Note 1**).
2. 8 wt./vol.% Polycaprolactone (PCL): Dissolve 0.4 g of PCL in 5 ml of 1,1,1,3,3,3-hexafluoro-2-propanol (HFIP) (*see* **Note 1**).
3. 18-gage needle.
4. T-shaped spinneret tube.
5. Syringe pumps.
6. Rotating collector.

2.2 Cross-Linking

1. 7 mM *N*-(3-Dimethylaminopropyl)-*N*′ ethylcarbodiimide hydrochloride (EDC) solution (cross-linking solution): Dissolve EDC in sterile 100% ethanol.
2. Sterile 70% ethanol.
3. Sterile phosphate-buffered saline (PBS).
4. Biopsy punch (8 mm).
5. Sterile forceps.
6. 94-mm dishes.

2.3 Culture of hiPSC-CMs

1. hiPSC-CMs (Fujifilm Cellular Dynamics, WI): Thaw and culture hiPSC-CMs as per the manufacturer's instructions on 1% gelatin-coated 6-well plates [11].
2. 50 μg/ml Fibronectin solution: Dissolve 50 μg of human fibronectin in 1 ml of PBS.
3. iCell cardiomyocyte maintenance medium (CMM, Fujifilm Cellular Dynamics, WI).
4. 0.25% trypsin–EDTA.
5. 94-mm tissue culture-treated dishes.
6. 6-well plates.
7. Sterile sponge (*see* **Note 2**).
8. Sterile N-Terface (Winfield Labs; *see* **Note 2**).
9. Autoclaved forceps and scissors.

2.4 Scanning Electron Microscopy

1. 4% paraformaldehyde (PFA) solution: Dissolve 4 g of PFA in 100 ml of PBS.
2. 50%, 70%, 80%, 95%, and 100% ethanol gradients (in PBS).
3. 25%, 50%, 75%, and 100% Hexamethyldisilazane (HDMS) gradients (in ethanol).
4. Carbon tape.

2.5 Confocal Microscopy

1. Gelatin-fluorescein solution: Dissolve 1 g of fluorescein in 100 ml of 12% gelatin solution.
2. PCL–rhodamine solution: Dissolve 1 g of rhodamine in 100 ml of 8% PCL solution.
3. Glass slides.
4. Coverslips.
5. Prolong diamond antifade.

3 Methods

3.1 Scaffold Preparation

1. Setting up a CoA spinneret: Insert an 18-G needle through a T-shaped spinneret tube to obtain a CoA spinneret setup. Manually wrap the 18-G needle with a thin wire to ensure that the inner needle is positioned at the center of the T-shaped spinneret and does not move vertically or laterally (*see* **Notes 3** and **4**). Figure 1 shows the schematic of the entire electrospinning setup (*see* **Note 5**).

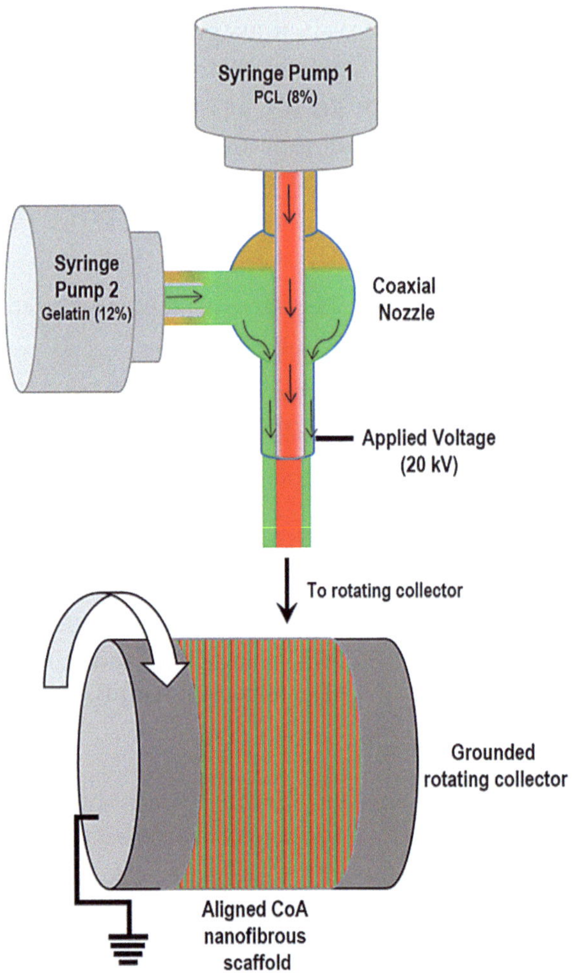

Fig. 1 Electrospinning setup for fabrication of aligned coaxial (CoA) PCL–gelatin nanofibrous scaffold. The 8% PCL solution (red) and 12% gelatin (green) were pumped at 1 ml/h and 4 ml/h, respectively, through the coaxial nozzle such that gelatin flows through the outer nozzle while PCL flows through the inner nozzle. A voltage of 20 kV is applied to the nozzle tip. The extruded nanofibers have an outer shell of gelatin and an inner core made of PCL. The fibers are collected onto a rotating collector to obtain aligned CoA nanofibrous scaffolds

2. Set up the syringe pumps to feed 12 wt./vol.% gelatin solution and 8 wt./vol.% PCL solution at flow rates of 4 ml/h and 1 ml/h, respectively (*see* **Note 6**).
3. Adjust the distance between the spinneret nozzle and the grounded rotating collector to 20 cm (*see* **Note 7**).
4. Maintain the co-axial spinneret nozzle at a voltage of 20 kV.
5. Collect the aligned co-axial nanofibers which start forming on the rotating collector, when the voltage is applied on the nozzle and the collector grounded.
6. Allow the collected aligned co-axial nanofibers to dry overnight in the chemical hood in order to remove the remnant solvent (*see* **Note 8**).

3.1.1 Support Protocol 1: Scanning Electron Microscopy of Scaffolds

The alignment of the nanofibers in the scaffold and the uniformity of the fibers can be analyzed by scanning electron microscopy (SEM, Fig. 2):

1. Cut a 0.5-cm × 0.5-cm piece of the nanofibrous scaffold.
2. Mount the piece onto a carbon tape and subject it to gold–palladium sputter coating as per the instrument manufacturer's instructions.
3. Image the coated scaffolds under a scanning electron microscope.

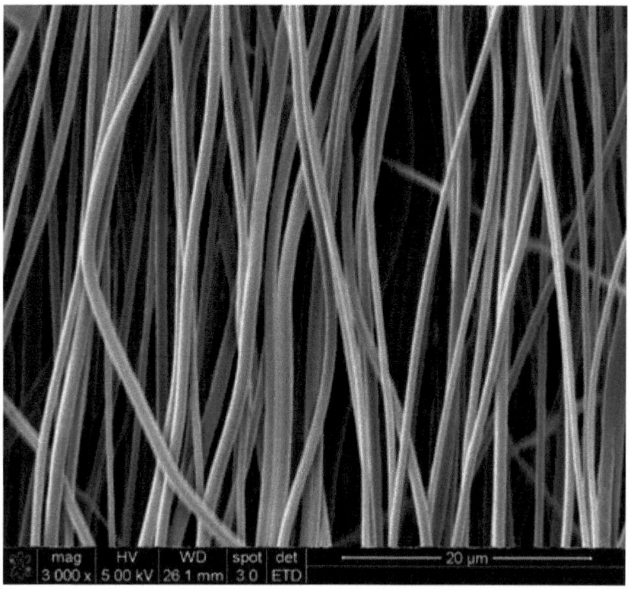

Fig. 2 Scanning electron microscopy (SEM) of coaxial aligned nanofibers: SEM image shows aligned (parallel) nanofibers within the scaffold. Scale bar: 20 μm

3.1.2 Support Protocol 2: Confocal Microscopy of Scaffolds

The co-axial nature of the nanofibers formed can be verified by confocal microscopy as previously reported [8]:

1. Use gelatin and PCL solutions containing fluorescein and rhodamine, respectively, for fabrication of the nanofibers as described above.
2. Cut small pieces of the nanofiber scaffold and follow the cross-linking process as mentioned in Subheading 3.2.
3. Place a small piece of the scaffold onto a glass slide.
4. Mount a glass coverslip on the scaffold with 1–2 drops of Prolong Diamond Antifade.
5. Image the nanofibers under a confocal microscope (*see* **Note 9**).

3.2 Cross-Linking and Sterilization

1. Using a biopsy punch, cut the aligned CoA nanofibrous scaffold into small circular pieces (Fig. 3).
2. Place the scaffold in a 94-mm dish containing cross-linking solution overnight at room temperature.
3. Wash the scaffold in 70% ethanol for 24 h to sterilize the scaffold.
4. Wash the scaffold thrice in sterile PBS in the biosafety cabinet.

3.3 Seeding hiPSC-CMs on an Aligned Nanofibrous Scaffold

3.3.1 Coating of Scaffold

1. Cut out 20-mm × 20-mm squares of the sterile N-Terface and place the pieces carefully in a 94-mm dish (Fig. 3).
2. Place one cross-linked 8-mm scaffold on the N-Terface carefully using sterile forceps (*see* **Note 10**).
3. Add 30 μl of fibronectin solution onto each scaffold. Avoid spilling over of the fibronectin out of the scaffold (*see* **Note 11**).
4. Incubate the scaffold at 37 °C for at least 1 h.

3.3.2 Seeding Cells on the Scaffold

1. Place a sterile sponge in a 94-mm dish and add 15 ml of PBS over it (Fig. 3). Once the sponge is completely soaked in PBS, aspirate out the PBS. Lightly press the sponge to ensure complete removal of PBS. Repeat this step twice.
2. Repeat **step 1** with CMM.
3. Transfer the scaffold along with the N-Terface onto the sponge as shown in Fig. 3.
4. Wash the hiPSC-CMs twice with PBS to remove any traces of medium in it.
5. Add 1 ml of 0.25% trypsin–EDTA (pre-warmed at 37 °C) per well of a 6-well plate.
6. Incubate the cells at 37 °C for 5–7 min (*see* **Note 12**).

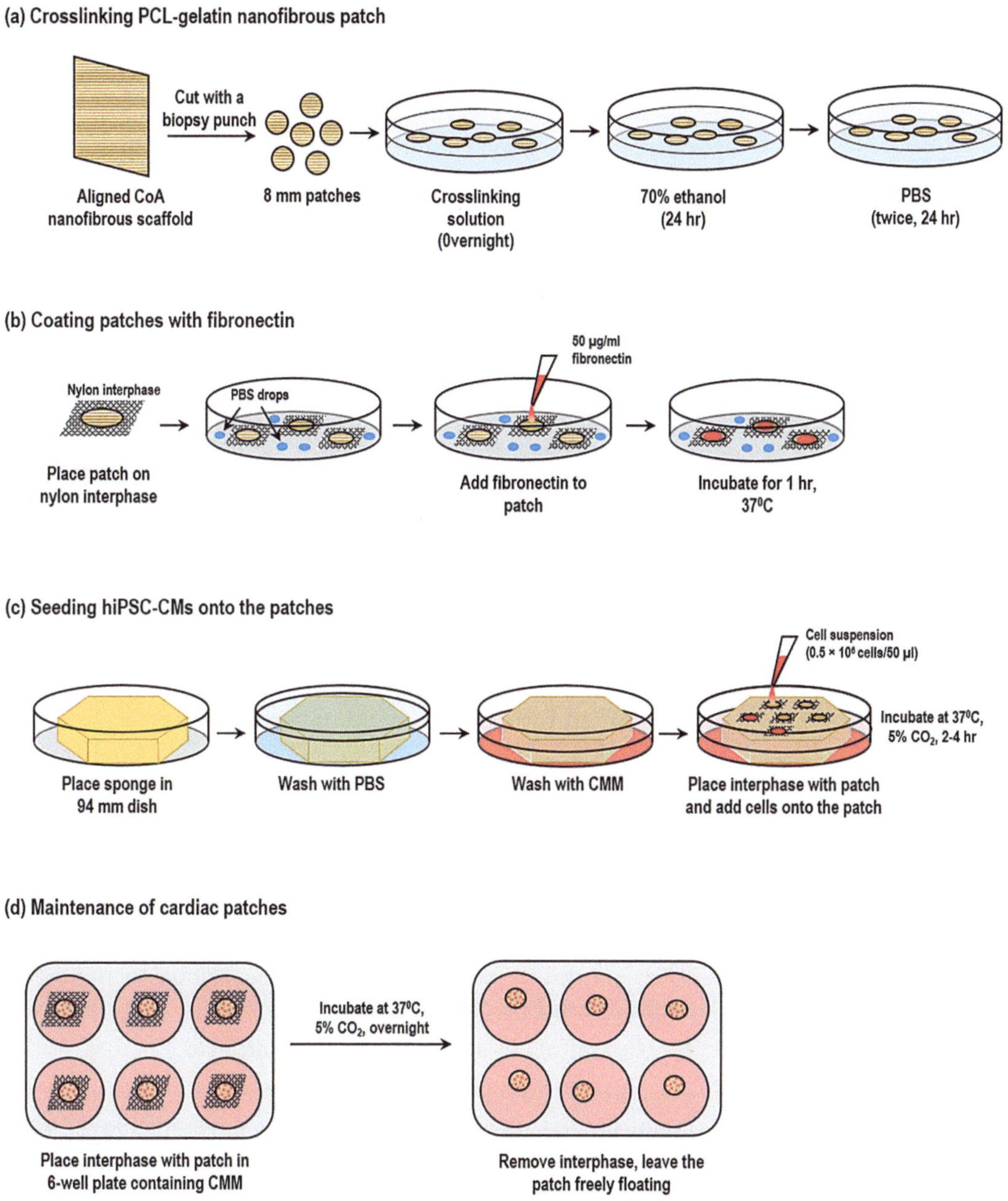

Fig. 3 Crosslinking, coating, and seeding hiPSC-CMs on the aligned nanofibrous scaffold. (**a**) The aligned nanofibrous scaffold obtained by electrospinning is cut into an 8-mm scaffold using a biopsy punch and cross-linked using 7 mM EDC solution. The cross-linked scaffold is sterilized by washing with 70% ethanol followed by PBS. (**b**) For coating, the cross-linked scaffold is placed on the N-Terface and kept in 94-mm dishes containing drops of PBS (to prevent evaporation). Thirty microliters of 50 μg/ml fibronectin is added onto the scaffold and incubated for 1 h at 37 °C. (**c**) For seeding hiPSC-CMs onto the scaffold, a sterile sponge is placed in a 94-mm dish and sequentially washed with PBS and cardiomyocyte maintenance medium (CMM). The fibronectin-coated scaffold along with the N-Terface is placed on the inoculation sponge, and cell suspension is added dropwise onto the scaffold. The dish is incubated for 2–4 h at 37 °C, 5% CO_2. (**d**) The scaffold along with the N-Terface is placed in a 6-well plate containing CMM and incubated overnight. The following day, the N-Terface is removed and scaffolds are left freely floating in CMM

7. Add equal volume of CMM to the cells to neutralize trypsin activity. Dislodge the cells from the cell culture plate by gentle pipetting (*see* **Note 13**).
8. Collect the cells into a 15-ml centrifuge tube. Centrifuge the cells at 300 × *g* for 10 min at RT.
9. Re-suspend the cell pellet in 5 ml of CMM and count the number of cells.
10. Centrifuge the cells again at 300 × *g* for 10 min at RT and re-suspend the cells in appropriate volume of CMM to obtain 0.5 million cells/50 μl (*see* **Note 14**).
11. Add 25 μl of cell suspension onto each scaffold. After 2–5 min, add 25 μl of cell suspension once again onto each scaffold (*see* **Note 15**).
12. Carefully place the dish in the incubator at 37 °C, 5% CO_2 for 4–6 h.
13. At the end of incubation, carefully pick the N-Terface along with the scaffold using forceps and transfer them into a 6-well plate containing 2 ml of CMM/well (*see* **Note 16**).
14. Incubate the scaffold at 37 °C, 5% CO_2 overnight.
15. The following day, carefully remove the N-Terface from under the scaffold without disturbing the orientation of the scaffold (*see* **Note 17**, Fig. 3).
16. Add an additional 1 ml of CMM per well of the 6-well plate.
17. Change the medium for the scaffold after every 48 h.

Support Protocol 3: Scanning Electron Microscopy of Cells

The hiPSC-CMs seeded onto the scaffold can be imaged using scanning electron microscopy (SEM, Fig. 4):

1. Wash the hiPSC-CMs cultured on the aligned nanofibrous scaffold with PBS (two times).
2. Fix the scaffold in 4% PFA for 20 min at RT.
3. Wash the scaffold twice in PBS.
4. Gradually dehydrate the scaffold using 50%, 70%, 80%, 95%, and 100% ethanol gradients. Incubate the scaffold for 20 min in each gradient solution.
5. Gradually dry the scaffold using 25%, 50%, 75%, and 100% HDMS gradients. Incubate the scaffold for 10 min in each solution. After incubation in 100% HDMS for 10 min, remove HDMS and allow it to evaporate completely in the hood before cutting the scaffold.
6. Make a cut on the scaffold (approximately 75% of the diameter).

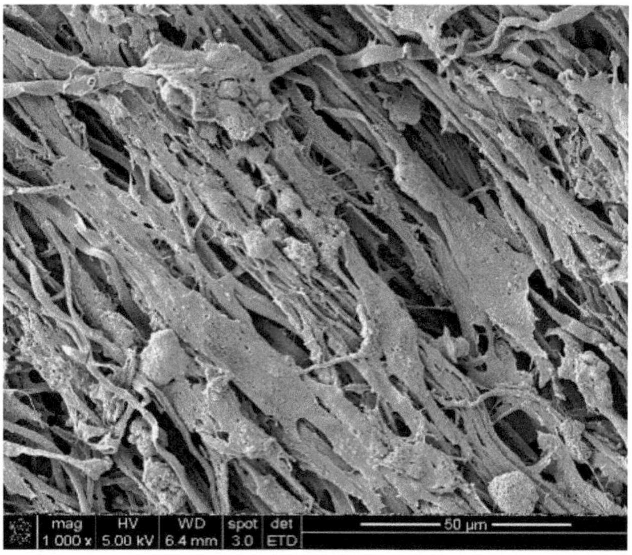

Fig. 4 Scanning electron microscope imaging of hiPSC-CMs cultured on aligned nanofibers: SEM image shows stoichiometric alignment of hiPSC-CMs in the direction of nanofibers within the cardiac scaffold. Scale bar: 50 μm

7. Mount the scaffold onto the carbon tape and fold the scaffold along the slit such that both the sides of the scaffold are exposed on the top for sputter coating.
8. Subject the scaffold to gold–palladium sputter coating as per the instrument manufacturer's instructions.
9. Image the coated scaffolds under a scanning electron microscope.

4 Notes

1. Cover the beaker to prevent evaporation of the solvent. Place the solution on a magnetic stirrer to ensure complete dissolution. Leave the PCL and gelatin solutions for 2 h and 48 h, respectively, for complete dissolution.
2. Sterilize the sponge and N-Terface using ethylene oxide gas or gamma irradiation prior to use in cell culture.
3. Make sure the 18-G needle (the inner tube) and the T-shaped spinneret tube (the outer one) are snugly fit and are not moving. This is very crucial to obtain co-axial nanofibers.
4. The inner needle acts as a tube which delivers the core (PCL) solution while the outer tube delivers the shell (Gelatin solution).

5. The electrospinning process is very sensitive to humidity. Make sure to maintain the setup between 35% and 40% relative humidity.
6. While using the syringe pump, make sure to select the appropriate syringe diameter and the flow rate.
7. Make sure the CoA nozzle is aligned in line with the rotating collector to maximize the nanofiber collection yield.
8. Make sure to remove all the remnant solvent by keeping the scaffold inside the chemical hood for at least 24 h.
9. Ensure crosslinking is performed before confocal imaging to stabilize the gelatin shell. Gelatin degrades quickly without crosslinking.
10. Ensure that the scaffold lies flat on the N-Terface and no crease appears on it to prevent improper coating and/or distribution of cells on the scaffold. In case the scaffold folds while placing, gently straighten it using blunt forceps. Be careful to avoid penetrating/tearing the scaffold.
11. Allow the scaffold to dry for 1–2 min before addition of fibronectin onto it. This would help retain the fibronectin on the scaffold and not flow out of it.
12. Monitor the cells every 3–4 min to observe the changes in their morphology. The duration of trypsinization can be extended till the cells start to detach from each other as well as from the dish. Gently rock the dish back and forth every few minutes to dislodge the cells. Do not tap the dish!
13. Avoid over pipetting.
14. The cell density to be plated is one million/cm^2. If scaffolds of different sizes are to be used, the cell numbers must be varied accordingly.
15. Add the medium dropwise onto the scaffold. Make sure the cell suspension is added evenly throughout the scaffold and not concentrated in one region. Add the second 25 μl after the first medium drop has been completely absorbed by the sponge. Otherwise, the cells may spread out of the scaffold.
16. Place the N-Terface such that it floats on the surface of the medium.
17. When removing the N-Terface, gently push it into the medium. When the scaffold comes off and floats freely, take out the N-Terface. Make sure that the scaffold does not fold, crease, or invert while doing so.

5 Troubleshooting

1. Low viability of hiPSC-CMs after trypsinization: Loss of cell viability may occur if the cells are over-trypsinized. Neutralize trypsin activity once the cells start to detach from the dish. Minimize the number of times the cells are pipetted to re-suspend the cell pellet.
2. Poor attachment of hiPSC-CMs onto the scaffold: This may occur in case of improper fibronectin coating. Do not allow the fibronectin drop to dry during coating. Additional drops of PBS may be placed in the dish to maintain humidity in the dish. If the problem persists, increase the concentration of fibronectin used for coating (e.g., 60–70 μg/ml) or leave the scaffolds on the sponge overnight to give more time of attachment. Also, while handling the scaffold, make sure that the tip of the forceps does not scrap off cells attached to the scaffold.
3. No contracting (beating) scaffolds: Normally, the contracting scaffolds are observed with 48–72 h of seeding the hiPSC-CMs onto the scaffold. In case no contractility is observed,
   (a) Ensure that the hiPSC-CMs show a synchronous contractility in gelatin-coated dishes before seeding them onto the scaffold. hiPSC-CMs may show some batch-to-batch variation in the initiation of synchronous contractility. However, the time may vary depending on the batch of hiPSC-CMs used.
   (b) If SEM analysis shows absence of cells, the drop of cell suspension must be added onto the scaffold. Make sure that the drop does not spread onto the sponge.
   (c) SEM analysis showed nonuniform distribution of cells: The cells need to distribute uniformly and form a syncytium for the scaffold to contract. During seeding, add the cells dropwise throughout the scaffold. However, if the problem persists, increase the seeding cell density to 1.5–two million/cm^2.

6 Anticipated Results

The synthesized aligned coaxial nanofibrous scaffold should show a parallel arrangement with each fiber showing a similar thickness (Fig. 2). Visible contacting cardiac scaffolds should be seen within 2–4 days of seeding the cells, and SEM analysis must show CMs aligned in the direction of the fibers (Fig. 4).

Acknowledgments

This work was supported by the National Institutes of Health (HL136232 MK), and OSU start-up funds to MK. We thank the Ohio State Campus Microscopy Imaging Facility for helping with SEM studies. These facilities are supported in part by grant P30 CA016058, National Cancer Institute, Bethesda, MD, USA.

References

1. Tang JN, Cores J, Huang K et al (2018) Concise review: is cardiac cell therapy dead? embarrassing trial outcomes and new directions for the future. Stem Cells Transl Med 7 (4):354–359. https://doi.org/10.1002/sctm.17-0196
2. Qasim M, Arunkumar P, Powell HM et al (2019) Current research trends and challenges in tissue engineering for mending broken hearts. Life Sci 229:233–250. https://doi.org/10.1016/j.lfs.2019.05.012
3. Gao L, Gregorich ZR, Zhu W et al (2018) Large cardiac muscle patches engineered from human induced-pluripotent stem cell-derived cardiac cells improve recovery from myocardial infarction in swine. Circulation 137 (16):1712–1730. https://doi.org/10.1161/CIRCULATIONAHA.117.030785
4. Menasche P, Vanneaux V, Hagege A et al (2018) Transplantation of human embryonic stem cell-derived cardiovascular progenitors for severe ischemic left ventricular dysfunction. J Am Coll Cardiol 71(4):429–438. https://doi.org/10.1016/j.jacc.2017.11.047
5. Rodrigues ICP, Kaasi A, Maciel Filho R et al (2018) Cardiac tissue engineering: current state-of-the-art materials, cells and tissue formation. Einstein (Sao Paulo) 16(3):eRB4538. https://doi.org/10.1590/S1679-45082018RB4538
6. Rane AA, Christman KL (2011) Biomaterials for the treatment of myocardial infarction: a 5-year update. J Am Coll Cardiol 58 (25):2615–2629. https://doi.org/10.1016/j.jacc.2011.11.001
7. Akhshabi S, Biazar E, Singh V et al (2018) The effect of the carbodiimide cross-linker on the structural and biocompatibility properties of collagen-chondroitin sulfate electrospun mat. Int J Nanomedicine 13:4405–4416. https://doi.org/10.2147/IJN.S165739
8. Blackstone BN, Drexler JW, Powell HM (2014) Tunable engineered skin mechanics via coaxial electrospun fiber core diameter. Tissue Eng Part A 20(19-20):2746–2755. https://doi.org/10.1089/ten.TEA.2013.0687
9. Li Y, Zhang D (2017) Artificial cardiac muscle with or without the use of scaffolds. Biomed Res Int 2017:8473465. https://doi.org/10.1155/2017/8473465
10. Blackstone BN, Hahn JM, McFarland KL et al (2018) Inflammatory response and biomechanical properties of coaxial scaffolds for engineered skin in vitro and post-grafting. Acta Biomater 80:247–257. https://doi.org/10.1016/j.actbio.2018.09.014
11. Kumar N, Dougherty JA, Manring HR et al (2019) Assessment of temporal functional changes and miRNA profiling of human iPSC-derived cardiomyocytes. Sci Rep 9 (1):13188. https://doi.org/10.1038/s41598-019-49653-5

Chapter 14

Generation of Myocardial Ischemic Wounds and Healing with Stem Cells

Daniela Rolph and Hiranmoy Das

Abstract

Myocardial ischemia is a common manifestation of cardiovascular diseases (CVD) that affects the health and lives of millions of people worldwide. While numerous treatment options exist that address cardiac damage after ischemic injury, none of these can repair damaged cardiac tissue. Stem cell-mediated therapy is an emerging approach for cardiac tissue regeneration that has shown promise in preclinical models and in clinical studies. However, much more research in this field must be carried out to bring effective stem cell therapies to clinical settings. This protocol discusses the methods for generation of an animal model of myocardial ischemia in a preclinical setting, expansion of viable hematopoietic stem cells on a nanofiber scaffold, and administration of cells into the ischemic animal to verify therapeutic efficacy.

Key words Myocardial ischemia, Athymic rat, Nanofiber-expanded $CD34^+$ stem cells

1 Introduction

Cardiovascular disease (CVD) and its complications are a major cause of morbidity and mortality around the world. Acute myocardial infarction is the most common manifestation of CVD. In the United States alone, about 790,000 people experience myocardial infarctions each year; of these, 580,000 are first-time events and 210,000 are recurrent events [1]. Myocardial ischemia leads to myocardial infarction, which can cause major damage to the heart. Myocardial ischemia is characterized by narrowing of the coronary arteries supplying the heart with blood. This causes reduced blood perfusion, which in turn leads to decreased oxygen availability and tissue damage. Because cardiomyocytes have a limited ability to perform anaerobic metabolism, constant perfusion with oxygen-rich blood is critical for their viability and function.

Pharmacologic and surgical interventions are currently used to mitigate cardiac damage after myocardial infarction; however, these treatment strategies alone are insufficient to repair damaged cardiac tissue. Heart transplantation is the only strategy in current clinical

Hiranmoy Das (ed.), *Wound Regeneration: Methods and Protocols*, Methods in Molecular Biology, vol. 2193,
https://doi.org/10.1007/978-1-0716-0845-6_14,

practice that can address ischemic heart failure after myocardial infarction; however, this treatment is costly and poses significant risks to the patient [2]. Stem cell-mediated regenerative approaches are currently under investigation as a potential treatment for myocardial damage. Infusion of hematopoietic stem and progenitor cells shows promise as a therapeutic approach for the treatment of numerous hematological disorders [3, 4]. This strategy is limited, however, by low recovery of donor stem cells, limited cell viability, and limited expansion techniques [5, 6].

Stem and progenitor cell transplantation has been shown to favorably impact tissue perfusion and contractile performance of cardiac cells in both preclinical animal studies [7] and in clinical trials [8]. Moreover, this strategy has been shown to promote neovascularization and myocyte formation in damaged cardiac tissue [9–11]. Our lab has extensive experience in the field of stem cell biology and has published work showing the therapeutic potential of human hematopoietic stem cells in various ischemic models [12, 13].

This protocol will demonstrate how stem cells can be used in an in vivo model of myocardial ischemia. It will first detail how stem cells can be expanded without negatively impacting viability or stem cell phenotype by growth on aminated nanofiber scaffolds [14]. Furthermore, the methodology for induction of myocardial ischemia in rats will be discussed, as will infusion of stem cells to treat damaged tissue.

2 Materials

All experiments should be conducted under sterile conditions in an approved biosafety level 2 (BSL2) laboratory. Reagents may be prepared at room temperature. All experiments with human samples must have Institutional Review Board (IRB) approval and all work carried out in animals must have Institutional Animal Care and Use Committee (IACUC) approval. Waste disposal should be performed according to standard regulations of biohazard disposal.

2.1 CD133+ Stem Cell Isolation and Expansion

1. Human umbilical cord blood.
2. Phosphate-buffered saline (PBS).
3. Ficoll-Paque PLUS density gradient media.
4. Centrifuge.
5. AutoMACS cell sorter (Miltenyi Biotec Inc.).
6. Anti-CD133 monoclonal antibody conjugated to magnetic beads (Miltenyi Biotec Inc.).
7. Polyether sulfone granules (MW = 55,000).
8. Dimethyl sulfoxide (DMSO).

9. Electrospinning syringe pump.
10. High-voltage power supply.
11. Flow cytometer.
12. StemSpan serum-free expansion medium (SFEM, Stem Cell Technologies) supplemented with 04 mg/ml low-density lipoprotein dissolved in NaCl (Athens Research and Technology Inc., USA), 100 ng/ml stem cell factor (SCF) dissolved in PBS, 100 ng/ml Flt3 dissolved in PBS, 50 ng/ml thyroid peroxidase (TPO) dissolved in PBS, and 20 ng/ml IL-3 dissolved in PBS.
13. Hanks' buffer supplemented with 2% FBS.
14. Nonenzymatic cell dissociation reagent.
15. Sterile BSL-2 biosafety cabinet.
16. Cell culture incubator (37 °C; 5% CO_2).
17. Sterile pipettes and tips.
18. Sterile microcentrifuge tubes and holding rack.

2.2 Animal Model of Myocardial Ischemia

1. Male athymic nude rats (6–8 weeks old) (Hsd:RH-rnu rats, Harlan Sprague–Dawley).
2. Sodium pentobarbital for injection.
3. Syringes and 25- and 27-gauge needles.

2.3 Histological Assessment

1. Paraformaldehyde (4%).
2. OCT compound.

3 Methods

3.1 CD133$^+$ Stem Cell Isolation

Fresh human umbilical cord blood should be collected after IRB approval and written donor consent (*see* **Notes 1** and **2**).

1. Collect cord blood (70–100 ml) in heparinized bags to prevent coagulation. After collection, dilute blood with PBS in a 1:1 ratio.
2. Carefully layer heparinized and diluted blood over Ficoll-Paque. Centrifuge for 30 min in a swinging bucket rotor at 1400 rpm (416 × *g*).
3. Following centrifugation, aspirate the upper layer and collect the mononuclear cell layer. Wash this layer three times with PBS.
4. Label mononuclear cells with anti-CD133 monoclonal antibody conjugated to magnetic beads. Isolate CD133$^+$ cells using an AutoMACS cell sorter system using double columns, using the manufacturer's protocol and reagents.

5. After separation, evaluate purity of isolated cells by flow cytometry.

3.2 Electrospinning, Surface Grafting, and Amination of PES Nanofiber Mesh

This procedure was developed based on the methods described by Chua et al. [15] and Joseph et al. [14]:

1. Dissolve polyether sulfone (PES) (MW = 55,000) granules in DMSO in a 20% solution.
2. Fill a plastic syringe fitted with a 27-gauge needle with the solution of PES in DMSO.
3. Use a syringe pump set to a feed rate of 0.3 ml/h to feed the polymer into the needle tip.
4. Fabricate the nanofibers in a sheet by electrospinning at 13 kV using a high-voltage power supply.
5. Wrap the sheet of electro spun nanofibers around a 15-mm-diameter glass coverslip, using medical-grade glue to adhere the fibers to the glass.

3.3 Ex Vivo CD133$^+$ Hematopoietic Cell Expansion (See Notes 1, 2, and 3)

1. Secure nanofiber mesh-wrapped coverslips to the bottoms of each of the 24 wells of a tissue culture plate using medical-grade glue. Ensure that the space between the bottom of the coverslips and the plate is sealed in order to prevent cell trapping in this space.
2. Seed 800 CD133$^+$ cells onto each scaffold in 600 μl of Stem-Span serum-free expansion medium supplemented with 0.04 mg/ml low-density lipoprotein, 100 ng/ml SCF, 100 ng/ml Flt3, 50 ng/ml TPO, and 20 ng/ml IL-3 (*see* **Note 4**).
3. Culture cells at 37 °C in an atmosphere containing 5% CO_2 without changing medium for 10 days.
4. Harvest cells after 10 days of expansion by washing once with a nonenzymatic cell dissociation solution and twice with Hanks' buffer supplemented with 2% FBS for 5–10 min.
5. Collect cell suspensions and centrifuge at 500 × *g* for 10 min.
6. Concentrated cells can be used for cell counting, flow cytometry, and further studies.

3.4 Animal Model of Myocardial Infarction (Fig. 1) (See Notes 3, 5, and 6)

1. Allow rats to acclimatize in the vivarium for at least 5 days.
2. Anesthetize rats with an intraperitoneal injection of sodium pentobarbital (50 mg/kg).
3. Induce myocardial infarct by surgically ligating the left anterior descending (LAD) coronary artery.
4. Allow 5 days of recovery before beginning cell transplantation experiments.

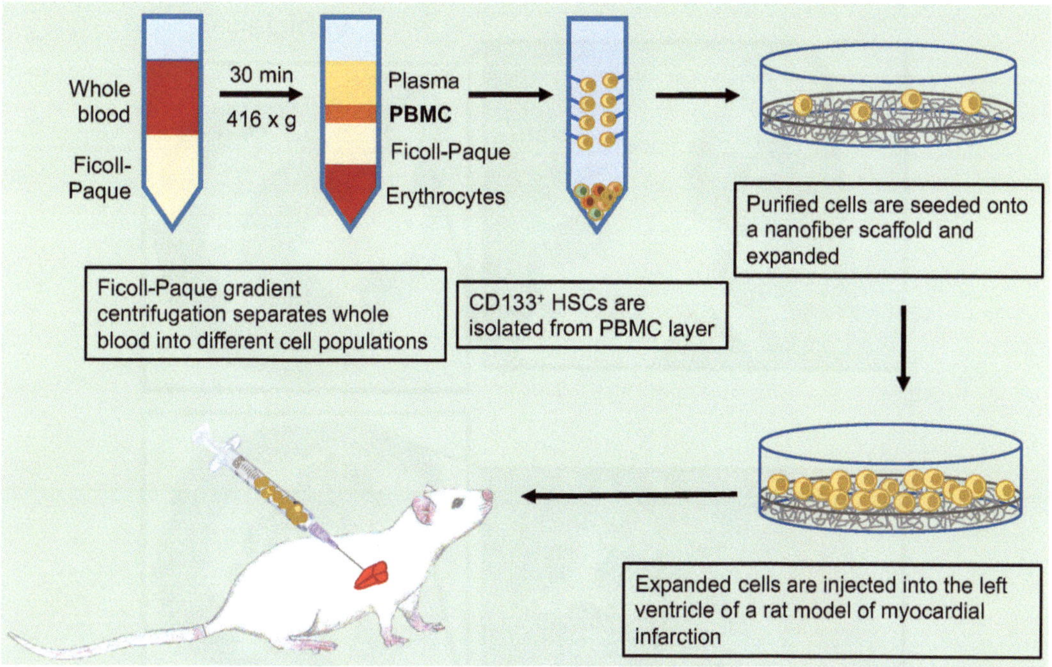

Fig. 1 Schematics of the major procedural steps in cell isolation, expansion, and transplantation. Whole blood is layered over Ficoll-Paque medium and centrifuged for 30 min at 416 × *g*. The PBMC layer is further purified using an AutoMACS cell sorter to isolate CD133$^+$ HSCs. These cells are seeded onto a nanofiber scaffold and expanded over the course of 10 days. The expanded cells are injected into the left ventricle of a rat model of myocardial infarction. Cell isolation and expansion is based on the methods reported by Das H. et al. [12]

3.5 Transplantation of Expanded Stem Cells (Fig. 2)

1. Five days after myocardial infarct, divide rats into groups ($n = 6$–9) to receive different treatment conditions.
2. Inject 5×10^5 cells suspended in 300 μl of media into the left ventricle (*see* **Notes 7** and **8**).
3. Sacrifice rats 6 weeks after intraventricular injection (*see* **Note 9**).

3.6 Histological Assessment of Transplanted Animals

1. Following euthanasia, slice hearts into 8 transverse sections from apex to base and fix in 4% paraformaldehyde.
2. Fix and embed half of the rat tissues in OCT compound and snap-freeze the other half in liquid nitrogen for fluorescence microscopy and alkaline phosphatase (ALP) staining.

4 Notes

1. All experiments require proper training in the techniques used therein.
2. Unless otherwise stated, all procedures should be carried out at room temperature under sterile conditions.

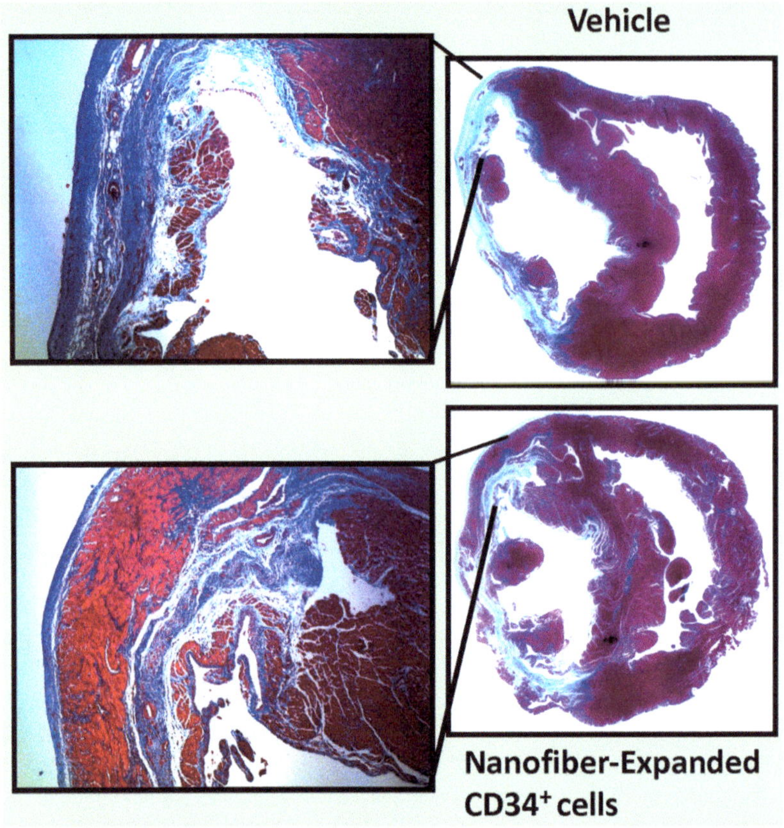

Fig. 2 Evaluation of total ischemic area using Masson's trichrome staining. Paraffin-blocked tissues from infarct zones of rat hearts were sectioned and stained with Masson's trichrome after six weeks of therapy. Decreased infarct size was observed in cell-treated versus vehicle-treated groups. Representative gross heart images are shown here with areas of interest in higher magnification. (Figure adapted from Das H. et al. [13])

3. Always check solutions for contamination before use, and evaluate cell conditions under a microscope. Wear appropriate personal protective equipment (PPE) (gloves and lab coat, etc.) when performing tissue culture experiments.
4. Resuspend cell pellets into a single cell suspension by firmly tapping the bottom of the centrifuge tube after discarding medium.
5. Always wear appropriate PPE (gloves, lab coat, face mask) when handling rodents.
6. Rats should be transported in closed cages; these should only be opened under a sterile hood in order to avoid exposing the animals to a nonsterile environment.
7. Before any injection, make sure there are no bubbles in the solution as these can cause air embolisms.

8. Confirm that the needle was inserted into the left ventricle by drawing blood back.
9. Monitor animals after all procedures.

Acknowledgments

This work was supported in part by National Institutes of Health grants, R01AR068279 (NIAMS), STTR 1R41EY024217 (NEI), and STTR 1R41AG057242 (NIA).

References

1. Benjamin EJ, Blaha MJ, Chiuve SE et al (2017) Heart disease and stroke Statistics-2017 update: a report from the American Heart Association. Circulation 135(10):e146–e603. https://doi.org/10.1161/CIR.0000000000000485
2. Ma T, Sun J, Zhao Z et al (2017) A brief review: adipose-derived stem cells and their therapeutic potential in cardiovascular diseases. Stem Cell Res Ther 8(1):124. https://doi.org/10.1186/s13287-017-0585-3
3. Gluckman EG, Roch VV, Chastang C (1997) Use of cord blood cells for banking and transplant. Oncologist 2(5):340–343
4. Tse W, Bunting KD, Laughlin MJ (2008) New insights into cord blood stem cell transplantation. Curr Opin Hematol 15(4):279–284
5. Segers VF, Lee RT (2008) Stem-cell therapy for cardiac disease. Nature 451 (7181):937–942
6. Daley GQ, Scadden DT (2008) Prospects for stem cell-based therapy. Cell 132(4):544–548
7. Asahara T, Masuda H, Takahashi T et al (1999) Bone marrow origin of endothelial progenitor cells responsible for postnatal vasculogenesis in physiological and pathological neovascularization. Circ Res 85(3):221–228. https://doi.org/10.1161/01.res.85.3.221
8. Leri A, Kajstura J, Anversa P et al (2008) Myocardial regeneration and stem cell repair. Curr Probl Cardiol 33(3):91–153. https://doi.org/10.1016/j.cpcardiol.2007.11.002
9. Kobayashi T, Hamano K, Li TS et al (2000) Enhancement of angiogenesis by the implantation of self bone marrow cells in a rat ischemic heart model. J Surg Res 89(2):189–195
10. Tomita S, Li RK, Weisel RD et al (1999) Autologous transplantation of bone marrow cells improves damaged heart function. Circulation 100(19 Suppl):II247–II256
11. Wu KH, Mo XM, Zhou B et al (2009) Cardiac potential of stem cells from whole human umbilical cord tissue. J Cell Biochem 107 (5):926–932. https://doi.org/10.1002/jcb.22193
12. Das H, Abdulhameed N, Joseph M et al (2009) Ex vivo nanofiber expansion and genetic modification of human cord blood-derived progenitor/stem cells enhances vasculogenesis. Cell Transplant 18(3):305–318. https://doi.org/10.3727/096368909788534870
13. Das H, George JC, Joseph M et al (2009) Stem cell therapy with overexpressed VEGF and PDGF genes improves cardiac function in a rat infarct model. PLoS One 4(10):e7325. https://doi.org/10.1371/journal.pone.0007325
14. Joseph M, Das M, Kanji S et al (2014) Retention of stemness and vasculogenic potential of human umbilical cord blood stem cells after repeated expansions on PES-nanofiber matrices. Biomaterials 35(30):8566–8575. https://doi.org/10.1016/j.biomaterials.2014.06.037
15. Chua KN, Chai C, Lee PC et al (2006) Surface-aminated electrospun nanofibers enhance adhesion and expansion of human umbilical cord blood hematopoietic stem/progenitor cells. Biomaterials 27(36):6043–6051. https://doi.org/10.1016/j.biomaterials.2006.06.017

Chapter 15

Corneal Repair Models in Mice: Epithelial/Mechanical Versus Stromal/Chemical Injuries

Peipei Pan and Matilda F. Chan

Abstract

The tissue response to injury is a complex process. The cornea is an excellent model for studying wound repair processes because of its simple anatomy, easy accessibility, and normal avascular state. Here, we describe two corneal repair models in mice: an epithelial/mechanical injury model and a stromal/chemical injury model. The two models induce different repair responses, and consequently enable the study of independent repair processes. Here, we describe how these two wound models may be used to study basic cellular and molecular mechanisms of corneal repair.

Key words Corneal injury, Fibrosis, Inflammation, Neovascularization, Chemokine, Macrophage, Neutrophil, MMP, CCL2/CCR2

1 Introduction

The cornea is the front part of the eye and has a critical role in the visual processing system by refracting light through the pupil and lens onto the retina. It also functions as an important protective barrier for the internal contents of the eye against various environmental challenges.

The cornea has a relatively simple anatomy and consists of three specialized cellular layers: epithelium, stroma, and endothelium. Distinct cell types allow each layer to have its own characteristic structure and function. The epithelium is the outermost cellular layer of the cornea and forms the first line of defense against pathogen invasion and external insults. It consists of 5–7 layers of stratified squamous nonkeratinized epithelial cells [1, 2]. The stroma is the thickest layer of the cornea and accounts for approximately 90% of the total corneal thickness [3]. It is mainly composed of water and has a sparse population of keratocytes within a highly ordered collagen-rich extracellular matrix (ECM). The stroma functions to provide mechanical strength and corneal transparency [2, 3]. The endothelium is the innermost, single-cell layer of cells.

Hiranmoy Das (ed.), *Wound Regeneration: Methods and Protocols*, Methods in Molecular Biology, vol. 2193,
https://doi.org/10.1007/978-1-0716-0845-6_15, © Springer Science+Business Media, LLC, part of Springer Nature 2021

The endothelial cells contain water channels whose roles are to maintain corneal water homeostasis [4].

Corneal integrity and transparency are critically important for good visual acuity. However, as the external surface of the eye, the cornea is susceptible to a variety of physical, chemical, and infectious insults. Injuries to the cornea may disrupt its normal function and anatomy and lead to visual impairment. Ocular morbidity as a result of corneal injuries can vary considerably, ranging from insignificant superficial wounds to more severe deeper injuries, which can lead to irreversible blindness. Depending on the extent and depth of injuries, the cornea wound-healing processes can vary largely and result in dramatically different repair outcomes.

Corneal wound healing is a complex and dynamic process that leads to the restoration of corneal tissue homeostasis. It involves a highly coordinated interaction of different types of cells, chemokines, growth factors, and ECM components [5–8]. In order to gain deeper insights into these interactions and processes, we have established two corneal repair models in mice: epithelial/mechanical and stromal/chemical injuries.

The epithelial/mechanical wound model injures the corneal epithelial layer and leaves the underlying corneal layers intact. This is a good injury model to study the repair processes of epithelial cell migration, proliferation, and reepithelialization. In this model, a defined area of the central cornea is marked with a trephine and an algerbrush burr is used to remove the epithelial cells within the mark. Fluorescein staining can be used to demarcate the area of the abraded epithelium and measure the rate of reepithelialization over time. We have used the epithelial/mechanical injury model to compare reepithelialization rates and inflammation between injured corneas of wild-type and *Mmp12*$^{-/-}$ mice [9]. This injury model allowed us to determine the role of MMP12 in promoting early repair processes following corneal epithelial injury by enhancing epithelial cell migration and neutrophil infiltration [9] (Fig. 1).

The stromal/chemical wound model injures the corneal epithelial and stromal layers. This deeper injury model induces an intense immune response and can be used to study the repair processes of corneal inflammation, neovascularization, and fibrosis. In this model, a circular filter paper is soaked in a chemical (e.g., 0.1 N NaOH) and then applied to the central cornea for a defined period of time. The strong immune response induces corneal neovascularization and the expression of fibrotic markers within a few days after injury. We have used the stromal/chemical wound model to compare the inflammatory, neovascular, and fibrotic responses between corneas of injured wild-type and *Mmp12*$^{-/-}$ mice [6, 7]. This deeper injury model allowed us to determine that MMP12 has a protective effect on corneal fibrosis during wound

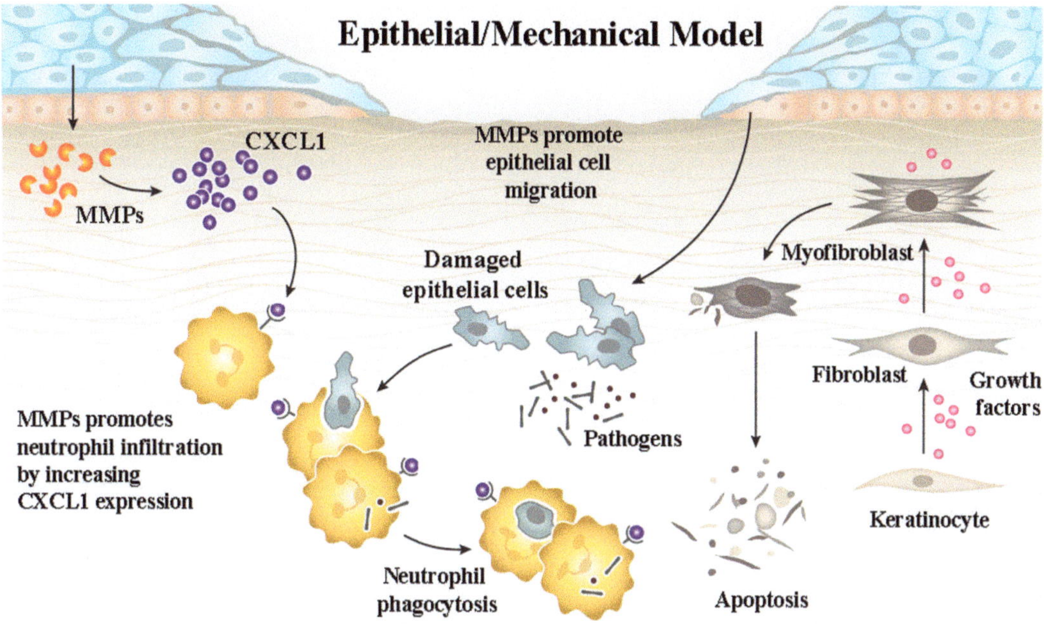

Fig. 1 Illustration of a corneal wound shows the cellular and molecular events after epithelial/mechanical injury. Corneal abrasion leads to increased expression of MMPs, in particular MMP12, which plays an essential role in promoting early corneal epithelial repair by increasing epithelial cell re-epithelialization and neutrophil infiltration. In addition, damaged epithelial cells produce growth factors which modulate keratinocyte phenotype and mechanical activity during epithelial cell migration

repair through its regulation of immune cell infiltration, angiogenesis, and CCL2/CCR2 signaling [6, 7] (Fig. 2).

Thus, we have established two corneal injury models in mice that enable the study of different corneal repair processes. We have used these models to determine how MMP12 regulates basic mechanisms involved in corneal reepithelialization, inflammation, neovascularization, and fibrosis. These corneal injury models can be used to further investigate and uncover additional basic repair mechanisms and are also suitable for clinical applications (e.g. preclinical drug safety and efficacy testing).

2 Materials

2.1 Animals and Reagents

2.1.1 Animals

It is important to age- and sex-match the comparative groups of mice because age and sex can influence the rates of corneal repair (*see* **Note 1**).

2.1.2 Reagents

All reagents may be prepared and stored at room temperature unless otherwise specified.

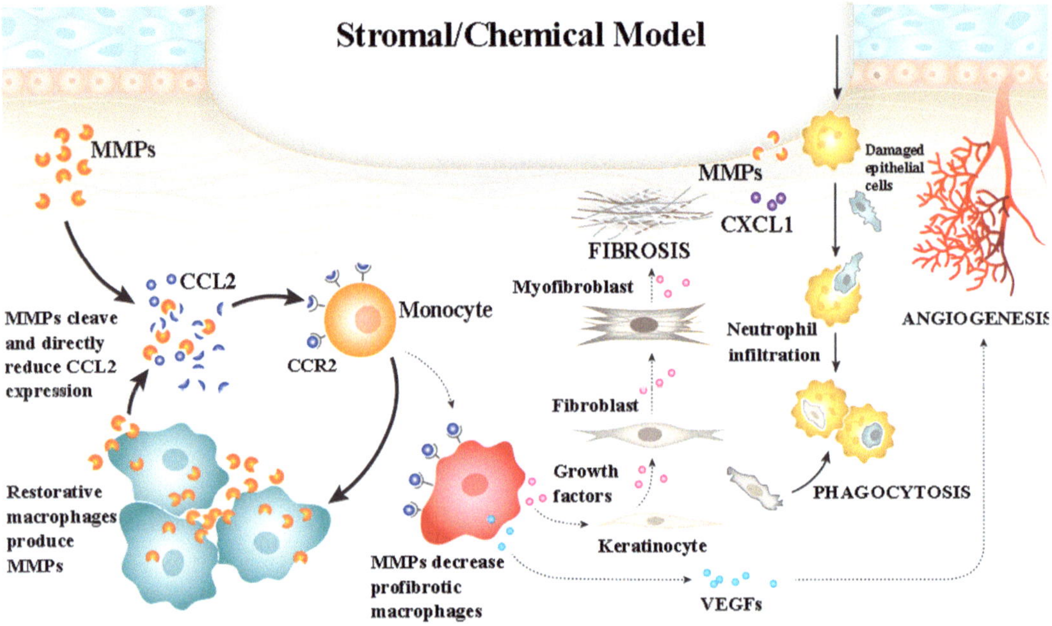

Fig. 2 Illustration of a corneal wound shows the cellular and molecular events after stromal/chemical injury. MMP12 expression is increased during corneal wound healing following chemical burns, which exerts a protective effect by inhibiting corneal fibrosis, neovascularization and inflammation through its direct downregulation of CCL2 and its primary receptor CCR2. VEGF, vascular endothelial growth factor

1. Sterile ultrapure water (H_2O).
2. Ethanol: 70% solution in water.
3. 1× sterile phosphate-buffered saline solution.
4. Sodium hydroxide (NaOH): 1.0 N (normal) solution. Prepare fresh 0.1 N NaOH by mixing 50 μl 1 N NaOH and 450 μl sterile H_2O to obtain a total volume of 500 μl.
5. Carprofen: Stock 50 mg/ml in PBS. Store at 4 °C. Prepare fresh on the day of the procedure by diluting with sterile 1× PBS into a final concentration of 0.5 mg/ml (final dosage 0.5 mg/kg body weight).
6. Isoflurane: Administered as the inhalant anesthesia using a nose cone system with a properly calibrated vaporizer (see below).
7. Proparacaine hydrochloride ophthalmic solution (0.5%; Bausch & Lomb, NDC: 24208-730-06).
8. Fluorescein sodium and benoxinate hydrochloride ophthalmic solution (0.25%/ 0.4%; Akorn, NDC: 17478-640-10). Prepare fresh on the day of the procedure by diluting (1:40) fluorescein solution in a 1.5 ml tube. Mix 2 μl of the fluorescein stock solution with 78 μl sterile 1 x PBS to obtain a total volume of 80 μl.

2.2 Supplies and Equipment

1. Reusable space gel heating pad (Fisher scientific, Cat. No.14-370-223).
2. Weck-cel cellulose eye spears (Medtronic, Cat. No. 0008680).
3. Algerbrush II with 0.5 mm Burr (Katena, Cat. No. K 2-4900).
4. Trephine with handle (1.5 mm; Beaver-visitec, Cat. No. 9748).
5. Stereomicroscope (for scratch/epithelial injury, need GFP filter; for both injuries, camera attachment is optional) (Leica, Cat. No. MZ16F).
6. Anesthesia machine with an induction chamber, nose cone attachment, and calibrated vaporizer (Summit Anesthesia Solutions, Cat. No. AS-01-0007).
7. Sterile, 1 ml, disposable syringe with needle (VWR, Cat. No. 76290-412).
8. Pipette (1 ml) with sterile aerosol tips (Fisher Scientific, Cat. No. 12111005).
9. Alcohol swabs—isopropyl alcohol 70% (BD, Cat. No. 326895).
10. Filter paper (Thermo Fisher Scientific, Cat. No. 09-795AA). Prepunched 2 mm filter paper discs using an ear punch into a sterile 1.5 ml Eppendorf tube.
11. Ear punch (Roboz, Cat. No. 65-9902).
12. Forceps (Dumont #5).
13. 15 ml conical sterile polypropylene centrifuge tube (Fisher Scientific, Cat. No. 1495970C)
14. 1.5 ml sterile Eppendorf tubes (Fisher Scientific, Cat. No. AM12400)
15. Timer (VWR, Cat. No. 62344-641).

3 Methods

Carry out all procedures at room temperature unless otherwise specified.

Ethical Statement: All procedures should be conducted in accordance with the tenets of the Declaration of Helsinki.

3.1 Procedure for Epithelial/ Mechanical Injury

1. Bench station set-up for surgical procedures (Fig. 3):
   (a) Use 70% ethanol to clean the bench station for the surgical procedures.
   (b) Place the anesthesia platform under the microscope objective and position the heating pad beneath it.

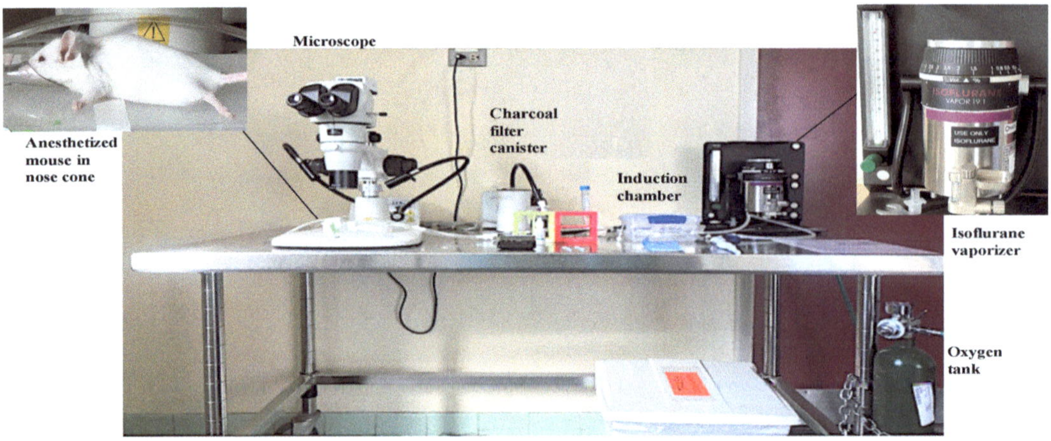

Fig. 3 Setup of the operating table for epithelial/mechanical and stromal/chemical injuries

(c) Affix the nose cone attachment durably on top of the anesthesia platform and connect the induction chamber to the isoflurane vaporizer.

(d) Place the proparacaine bottle, fluorescein solution, Algerbrush, trephine, and Weck-cel spears on the bench station.

(e) Add isoflurane to the anesthesia machine and connect it to the oxygen tank so that the system delivers a blended mixture of oxygen and isoflurane into the induction chamber for mouse anesthesia.

(f) Connect the induction chamber to a charcoal filter canister to capture waste anesthetic gases.

(g) Clean the Algerbrush burr using a 70% isopropyl alcohol swab prior to use.

2. Inject Carprofen intraperitoneally into each mouse based on its body weight to provide postsurgical pain control. Use the following equation to calculate the injection volume (*see* **Note 2**).

3. Open the gas supply to the induction chamber. Anesthetize a single mouse by placing it in the isoflurane induction chamber, closing the lid, and then adjusting the oxygen flow to 0.9 l/min and the isoflurane vaporizer to 1–2%. Once shallow breathing is noted, quickly transfer the mouse to the anesthesia platform under the microscope and fit the mouse into the nose cone. Divert the gas supply from the induction chamber to the nose cone. Check for proper mouse anesthesia using paw pinches. Once the mouse is fully anesthetized, position the mouse's head on the platform so that the eye to be injured is facing up toward the microscope objective.

4. Place one drop of Proparacaine on the mouse cornea and wait for 30 s.
5. Use a Weck-cel eye spear to gently dry off the cornea by dabbing both corners of the eye.
6. Use one hand to gently apply periocular pressure to proptose the eye. (Optional step: Take a bright-field image of the cornea prior to injury).
7. While looking through the stereo microscope, use the other hand to gently mark the central cornea with a trephine and turn it to approximately three clock hours to make a ring-shaped circle. Attempt to make a single mark.
8. Turn on the Algerbrush and make an epithelial defect in the center cornea. The remaining epithelium within the trephine mark can be further removed using the rotating burr. Alternatively, the remaining epithelium can be manually removed with the Algerbrush off and using its dull blades to extend the epithelial defect to the trephine mark.
9. Apply 20 μl of fluorescein solution to the injured cornea. Change to the GFP filter on the microscope to confirm the epithelial injury. (Optional step: take a picture of the epithelial defect using the GFP filter; Fig. 4).
10. Place another drop of Proparacaine solution onto the injured cornea to ease postsurgical pain.
11. Remove mouse from the nose cone and place it onto a heating pad in a recovery cage. Allow the mouse to awaken unperturbed. Monitor the mouse for signs of postsurgical pain and eye infections.

3.2 Procedure for Stromal/Chemical Injury

1. Bench station set-up for surgical procedures:
   (a) Use 70% ethanol to clean the bench station for the surgical procedures.
   (b) Place the anesthesia platform under the microscope objective and position the heating pad beneath it.
   (c) Affix the nose cone attachment durably on top of the anesthesia platform and connect the induction chamber to the isoflurane vaporizer.
   (d) Place the Proparacaine bottle, fluorescein solution, prepunched 2 mm filter discs, 0.1 N NaOH solution, forceps, and Weck-cel spears on the bench station. Have 500 μl 1× PBS ready in a pipette.
   (e) Add isoflurane to the anesthesia machine and connect it to the oxygen tank so that the system delivers a blended mixture of oxygen and isoflurane into the induction chamber for mouse anesthesia.

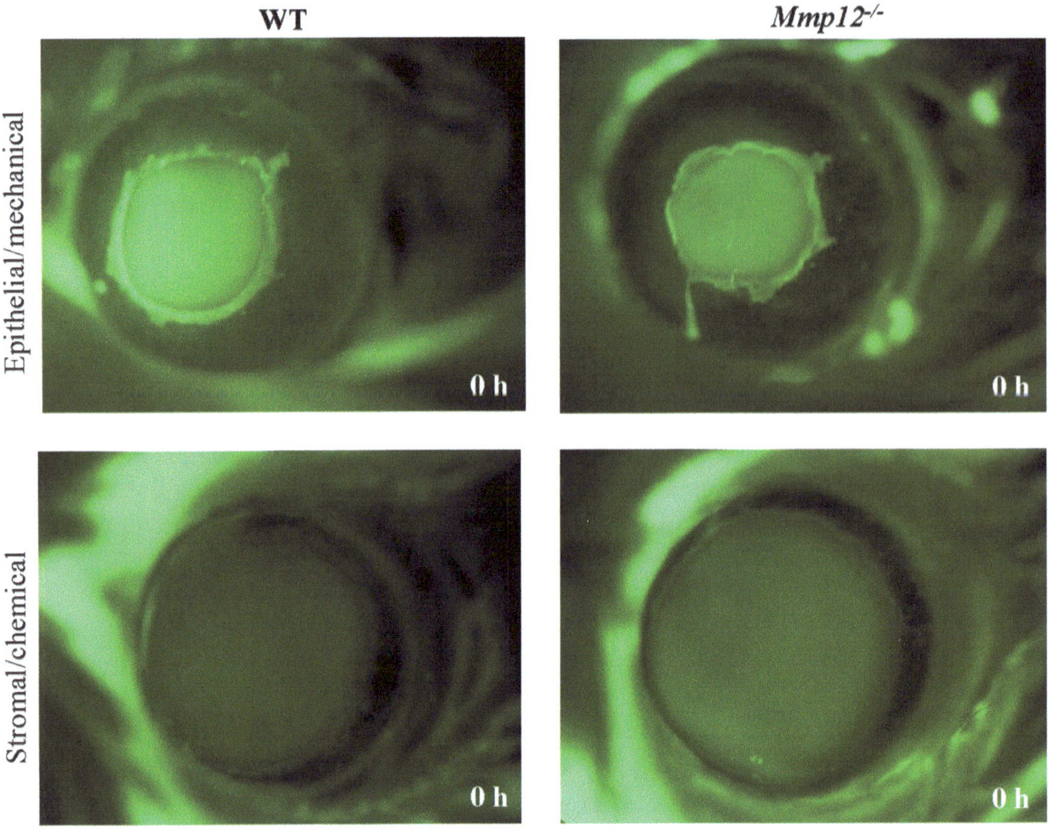

Fig. 4 Fluorescein staining of cornea reveal the epithelial defects of the center cornea after epithelial/mechanical and stromal/chemical injuries in WT and Mmp12$^{-/-}$ mice

(f) Connect the induction chamber to a charcoal filter canister to capture waste anesthetic gases.

(g) Clean the forceps tips using a 70% isopropyl alcohol swab prior to use.

(h) Set the timer for both 10 s and 30 s.

2. Inject Carprofen intraperitoneally into each mouse based on its body weight to provide postsurgical pain control. Use the following equation to calculate the injection volume (*see* the Note).

3. Open the gas supply to the induction chamber. Anesthetize a single mouse by placing it in the isoflurane induction chamber, closing the lid, and then adjusting the oxygen flow to 0.9 l/min and the isoflurane vaporizer to 1–2%. Once shallow breathing is noted, quickly transfer the mouse to the anesthesia platform under the microscope and fit the mouse into the nose cone. Divert the gas supply from the induction chamber to the nose cone. Check for proper mouse anesthesia using paw pinches.

Once the mouse is fully anesthetized, position the mouse's head on the platform so that the eye to be injured is facing up toward the microscope objective.

4. Place one drop of Proparacaine on the mouse cornea and wait for 30 s.
5. Use a Weck-cel eye spear to gently dry off the cornea by dabbing both corners of the eye.
6. Use the forceps to obtain and hold the edge of a filter paper. Fully immerse the filter paper into the 0.1 N NaOH solution for 10 s.
7. Use one hand to proptose the mouse eye by gently applying periocular pressure (Optional step: Take a bright-field image of the cornea prior to injury).
8. While looking through the stereo microscope, use forceps in the other hand to place the NaOH-soaked filter paper onto the center cornea for exactly 30 s. Use the forceps to gently press down on the filter paper to ensure that the entire disc makes contact with the cornea.
9. After 30 s, use the forceps to remove the filter paper disc from the cornea.
10. Quickly flush the eye with 500 μl 1× PBS to remove residual NaOH.
11. Gently dry off the ocular surface with Weck-cel spears.
12. Apply 20 μl of the fluorescein solution to the injured cornea. Change to the GFP filter on the microscope to confirm and assess the corneal injury. (Optional step: take a picture of the epithelial defect using the GFP filter; Fig. 4).
13. Place another drop of Proparacaine solution onto the injured cornea to ease postsurgical pain.
14. Remove the mouse from the nose cone and place it onto a heating pad in a recovery cage. Allow the mouse to awaken unperturbed. Monitor the mouse for signs of postsurgical pain and eye infections.

4 Notes

1. The use of littermate, age-matched, and sex-matched mice is recommended because these variables can influence the corneal repair process [10].
2. The following equation is used to calculate the exact volume of 0.5 mg/ml diluted Carprofen administered intraperitoneally per mouse at a final dosage of 5 mg/kg of body weight [11]:

$$\frac{\text{g of mouse} \times 0.005 \text{ mg Carprofen/g of body weight}}{0.5 \text{ mg/ml diluted Carprofen solution}}$$

Acknowledgments

We thank Suling Wang for assistance with preparing and illustrating the figures of the corneal wound models. This work was supported by grants from the National Institutes of Health (R01 EY022739 to M.F.C; NIH-NEI EY002162 - Core Grant for Vision Research), Research to Prevent Blindness (RPB Physician-Scientist Award to MFC; RPB Unrestricted Grant to the UCSF Department of Ophthalmology), and That Man May See.

References

1. Chi C, Trinkaus-Randall V (2013) New insights in wound response and repair of epithelium. J Cell Physiol 228(5):925–929. https://doi.org/10.1002/jcp.24268
2. Sridhar MS (2018) Anatomy of cornea and ocular surface. Indian J Ophthalmol 66 (2):190–194. https://doi.org/10.4103/ijo.IJO_646_17
3. Hassell JR, Birk DE (2010) The molecular basis of corneal transparency. Exp Eye Res 91 (3):326–335. https://doi.org/10.1016/j.exer.2010.06.021
4. Bonanno JA (2012) Molecular mechanisms underlying the corneal endothelial pump. Exp Eye Res 95(1):2–7. https://doi.org/10.1016/j.exer.2011.06.004
5. Eming SA, Wynn TA, Martin P (2017) Inflammation and metabolism in tissue repair and regeneration. Science 356(6342):1026–1030. https://doi.org/10.1126/science.aam7928
6. Chan MF, Li J, Bertrand A et al (2013) Protective effects of matrix metalloproteinase-12 following corneal injury. J Cell Sci 126 (Pt 17):3948–3960. https://doi.org/10.1242/jcs.128033
7. Wolf M, Clay SM, Zheng S et al (2019) MMP12 inhibits corneal neovascularization and inflammation through regulation of CCL2. Sci Rep 9(1):11579. https://doi.org/10.1038/s41598-019-47831-z
8. Ljubimov AV, Saghizadeh M (2015) Progress in corneal wound healing. Prog Retin Eye Res 49:17–45. https://doi.org/10.1016/j.preteyeres.2015.07.002
9. Wolf M, Maltseva I, Clay SM et al (2017) Effects of MMP12 on cell motility and inflammation during corneal epithelial repair. Exp Eye Res 160:11–20. https://doi.org/10.1016/j.exer.2017.04.007
10. Elliot S, Wikramanayake TC, Jozic I, Tomic-Canic M (2018) A Modeling Conundrum: Murine Models for Cutaneous Wound Healing. J Invest Dermatol 138:736–740. https://doi.org/10.1016/j.jid.2017.12.001
11. Kendall LV et al (2014) Pharmacokinetics of sustained-release analgesics in mice. J Am Assoc Lab Anim Sci 53:478–484

Chapter 16

Topical Adoptive Transfer of Plasmacytoid Dendritic Cells for Corneal Wound Healing

Arsia Jamali, Brendan M. Kenyon, Gustavo Ortiz, Betul N. Bayraktutar, Victor G. Sendra, and Pedram Hamrah

Abstract

Plasmacytoid dendritic cells (pDCs) are crucial for corneal homeostasis through secretion of various anti-angiogenic molecules and growth factors. Due to its avascular nature, only a limited number of adoptively transferred cells home to the cornea, when administered systemically. In addition, local adoptive transfer of cells poses several challenges and the clinical application of commonly used techniques is limited. Herein, we detail a novel approach for local adoptive transfer of pDCs to the cornea for the treatment of corneal wounds. This approach utilizes a commonly used fibrin sealant as a means of transferring previously isolated cells locally on the cornea. The technique is simple, reproducible, and is accompanied with successful transfer and integration of a substantial number of the cells to the cornea. Application of this approach to transfer pDCs promotes corneal wound healing. Furthermore, this technique can be applied for adoptive transfer of any cell of interest to the cornea.

Key words Plasmacytoid dendritic cells, Cornea, Wound, Adoptive transfer, Healing

1 Introduction

Plasmacytoid dendritic cells (pDCs) are a subtype of immune cells, which are involved in immune responses, ranging from antiviral immunity to the induction of tolerance [1–4]. Recently, it has been demonstrated that pDCs exhibit anti-inflammatory properties in the cornea during both viral infections and sterile inflammation [5–7]. Further, pDCs play a critical role in maintaining corneal homeostasis by preserving corneal angiogenic privilege and maintaining corneal innervation [8, 9]. Moreover, pDCs secrete multiple growth factors necessary for tissue regeneration following corneal insults [9, 10].

Arsia Jamali and Brendan M. Kenyon contributed equally to the work.

Hiranmoy Das (ed.), *Wound Regeneration: Methods and Protocols*, Methods in Molecular Biology, vol. 2193,
https://doi.org/10.1007/978-1-0716-0845-6_16,

A variety of approaches have previously been tested for adoptive transfer of cells of interest for the treatment of corneal diseases or for studying on the role of cells of interest in disease pathogenesis. The transferred cells have included stem cells, myeloid-derived suppressor cells, and subpopulations of T cells, such as γδ T cells, regulatory T cells, and in vitro-stimulated T-cell lines [11–17]. Systemic administration of the cells via intraperitoneal or intravenous injections is less favored [11–14], since considering the avascular nature of the cornea, systemic transfer of cells results in their distribution to multiple predominantly nonocular sites, such as the spleen and lymph nodes. In fact, via systemic approaches, only a minor fraction of transferred cells have been shown to home to the cornea [14–16]. This broad distribution of cells via systemic approaches increases the potential off-site adverse effects, and increases the number of cells required for adoptive transfer [14–16], making the transfer of rare cell types difficult and challenging the clinical feasibility of these methods. For local adoptive transfer of cells, subconjunctival and intrastromal injections have previously been employed [17, 18]. Considering that the conjunctiva contains a substantial density of lymphatic and blood vessels [19], subconjunctival transfer of cells may also lead to systemic distribution of cells [18]. The clinical application of intrastromal injections is limited due to its invasive nature and potential complications, such as corneal perforation [20–22].

Herein, we present a protocol of local adoptive transfer of pDCs to the cornea, utilizing commercially available fibrin sealants to promote corneal wound healing [7, 10, 23]. This approach has multiple advantages over other currently utilized techniques of adoptive transfer of cells, including its high efficacy, relatively simplicity, and high clinically feasibility. Further, this technique may overcome the above-mentioned shortcomings of comparable techniques, as the cells are locally transferred to the cornea and thus allow adoptive transfer of relatively low number of cells with superior efficacy of transfer. Notably, a significant fraction of the transferred cells can be observed within the cornea within hours and for up to at least 4 weeks after adoptive transfer. Furthermore, in addition to treatment of corneal wounds, this approach of direct topical adoptive transfer can be utilized to enrich corneas with any particular cells of interest to study their contribution in pathogenesis and/or to study their efficacy in treating a variety of corneal conditions, such as mechanical and chemical traumas, dry eye disease, infectious keratitis, neovascularization, transplantation, and nerve-mediated diseases. A potential limitation of this technique is the necessity of a temporary corneal epithelial curettage in order to allow the cells to integrate into the corneal stroma. Although for the purposes of adoptive transfer of cells, it is sufficient to establish a minimal epithelial debridement, which generally heals within 24–48 h in wild-type mice [24, 25], it may accompany corneal nerve damage that may alter or augment the pathophysiological processes underlying the aforementioned corneal diseases.

2 Materials

Unless otherwise noted, all buffers should be stored at 2–8 °C and kept on ice during the process. Prepare all buffers and solutions in advance, except for fibrin sealant, which should be prepared fresh at the time of use. Carefully follow all waste disposal regulations when disposing waste materials.

2.1 Donor Tissue Collection

1. Sterile surgical tools for dissection: two fine forceps and dissecting scissors.
2. Seventy percent ethanol for disinfection; store at room temperature.
3. Sterile ice-cold phosphate-buffered saline (PBS, calcium and magnesium free; *see* **Note 1**).
4. A six-well cell culture plate (*see* **Note 2**).

2.2 Tissue Processing and Single Cell Preparation

1. An 8–10-week-old C57BL/6J wild-type mouse (*see* **Notes 3** and **4**).
2. Sterile 1-mL syringes for mechanical dissociation of spleen(s).
3. Conical tubes of 50 mL.
4. Cell strainers of 40 μm.
5. Sterile ice-cold PBS.
6. Ammonium-Chloride-Potassium (ACK) red blood cell (RBC) lysis buffer: To prepare 1 L of ACK RBC lysis buffer, add 800 mL of dH_2O to a suitable beaker or flask. Add the following components: 8.29 g of NH_4Cl (0.15 M), 1 g of $KHCO_3$ (10 mM), and 37.2 mg of Na_2EDTA (0.1 mM). Add a stir bar and place on a magnetic stir plate until completely dissolved. Adjust the pH of the solution to 7.2–7.4 with HCl or NaOH. Add dH_2O to a final volume of 1 L. Filter the solution in a cell culture hood in a sterile fashion. Store the buffer at 2–8 °C (*see* **Note 5**).

2.3 Cell Staining and Sorting

1. Fluorescence-activated cell sorting (FACS) buffer: The following reagents should be added to 500 mL of calcium- and magnesium-free PBS to achieve the indicated concentrations: 146.1 mg of EDTA (1 mM) and 5 g of bovine serum albumin (1%). Sterile filter the solution and store at 2–8 °C (*see* **Note 6**).
2. Sterile FACS tubes.
3. Anti-mouse CD16/CD32 Fc block (for instance, BioXCell, West Lebanon, NH, USA or similar commercially available

products by other companies); store at −20 °C or colder prior to thawing for the experiment.

4. A viability marker; store as recommended by the manufacturer (*see* **Note 7**).
5. Fluorophore-conjugated antibodies for staining pDCs (*see* **Note 8**) including anti-PDCA-1 (CD317, BST2), anti-CD45R/B220, and anti-Siglec-H, as well as their respective isotype controls (*see* **Notes 9** and **10**).

2.4 Cornea Wounding and Local Adoptive Transfer of Plasmacytoid Dendritic Cells

1. Wild-type C57BL/6J mice (*see* **Note 11**).
2. Institutional animal care and use committee (IACUC)-approved anesthetic (for instance ketamine and xylazine mixture) is required for anesthetizing the animals for corneal wounding and adoptive transfer of pDCs.
3. Proparacaine hydrochloride ophthalmic solution (0.5%; Alcon, Fort Worth, Texas, USA) to numb the cornea prior to epithelial debridement.
4. Ophthalmic artificial tears to prevent desiccation of the contralateral cornea; store at room temperature.
5. Neomycin, polymyxin B, and bacitracin triple antibiotic ophthalmic ointment; store at room temperature.
6. Surgical instruments for debridement of the corneal epithelium: a curved forceps, 2 mm trephine, and Algerbrush II corneal rust ring with 0.5 mm burr (The Alger Companies, Lago Vista, Texas, USA).
7. Fibrin sealant (e.g., TISSEEL® [Baxter International, Deerfield, IL, USA]) composed of (1) sealer protein solution and (2) thrombin solution for adoptive transfer of pDCs. Fibrin sealant should be stored at −20 °C or colder until immediately before use.
8. Heating lamp.

2.5 Outcome Assessments

1. IACUC-approved anesthetic (for instance, ketamine and xylazine mixture) for anesthetizing the animals for proper assessment of corneal wounds.
2. Slit-lamp biomicroscope equipped with a camera adaptor.
3. Fluorescein ophthalmic solution (0.25%; *see* **Note 12**).
4. Normal saline solution.
5. Cotton tip applicator.
6. Heating lamp.

3 Methods

3.1 Donor Tissue Collection

1. Add 3 mL of ice-cold PBS to one well of a six-well cell culture plate (*see* **Note 13**).
2. Euthanize an 8–10-week-old C57BL/6 wild-type mouse according to IACUC-approved procedure (*see* **Notes 3** and **4**).
3. Spray 70% ethanol on the skin of the abdomen and chest to disinfect the animal.
4. Place the animal on its right flank, exposing the left flank. Promptly incise the skin, fascia, and peritoneum of the left upper quadrant of the abdomen and left flank of the mouse to explore the spleen. By cutting the connective tissues, dissect and collect the spleen in a well of a six-well cell culture plate (containing 3 mL of ice-cold PBS, prepared in the **step 1**).

3.2 Tissue Processing and Single Cell Preparation

1. Place a cell strainer on a 50 mL conical tube and set the spleen on the cell restrainer. Gently dissociate the spleen mechanically using the round bottom of a plunger of a sterile 1 mL syringe.
2. Once the tissue is disrupted, wash the strainer with 5–7 mL of ice-cold PBS.
3. Discard the cell strainer and centrifuge the flow through at 1200 rpm for 5 min, at 2–8 °C.
4. Aspirate the supernatant and resuspend the pellet in 8 mL of ice-cold ACK RBC lysis buffer and incubate the samples in the lysis buffer for 1 min, at room temperature.
5. After incubation, add at least 12 mL of ice-cold PBS and centrifuge at 1200 rpm for 5 min, at 2–8 °C (*see* **Note 14**).
6. Aspirate the supernatant and repeat washing with 20 mL of ice-cold PBS if necessary (optional; *see* **Note 15**).

3.3 Cell Staining and Sorting

1. Prepare 1.5 mL of blocking buffer by adding 15 μL (1:100) of antimouse CD16/CD32 Fc block to the FACS buffer (*see* **Note 16**) and a viability marker per manufacturer's instructions (*see* **Note** 7).
2. After completing centrifugation, aspirate the supernatant and resuspend the pellet in 1 mL of blocking buffer and incubate for 15 min, at 2–8 °C (*see* **Note 17**).
3. Divide the sample into two FACS tubes: (A) 100 μL of the sample for nonstained cell controls, as well as isotype controls, single stained controls, and fluorescence minus one controls for cell sorting, and (B) 900 μL of the sample for staining for pDCs with PDCA-1, CD45R/B220, and Siglec-H for actual sorting of pDCs.

4. Stain the samples and controls with fluorophore-conjugated antibodies and their respective isotype controls (all 1:100) as follows:
   (a) Add 700 μL of the blocking buffer to the cells in the control FACS tube (containing 100 μL of the original tube) and divide them into eight FACS tubes. Add 1 μL of fluorophore-conjugated antibodies or their isotype controls to prepare appropriate controls:
      - Unstained cells.
      - Isotype control for PDCA-1.
      - Isotype control for CD45R/B220.
      - Isotype control for Siglec-H.
      - Anti-PDCA-1.
      - Anti-CD45R/B220.
      - Anti-Siglec-H.
      - Anti-PDCA-1, anti-CD45R/B220, and Isotype control for Siglec-H.
   (b) Add 9 μL of each fluorophore-conjugated antibody against PDCA-1, CD45R/B220, and Siglec-H to the FACS tube containing 900 μL of the original cells.
   (c) Incubate the samples for 30–40 min, at 2–8 °C on a shaker with minimal exposure to light.
5. Add 3–4 mL of FACS buffer to each sample for washing, and centrifuge at 1200 rpm for 5 min, at 2–8 °C.
6. Resuspend the cells in at least 1 mL of FACS buffer.
7. Filter the cells through a 40 μm cell strainer (optional; *see* **Note 18**).
8. Sort pDCs as live PDCA-1$^+$ CD45R/B220$^+$ Siglec-H$^+$ cells via a flow cytometric sorter (*see* **Notes 19** and **20**; Fig. 1).

3.4 Cornea Wounding and Local Adoptive Transfer of Plasmacytoid Dendritic Cells

1. Wash the sorted pDCs with ice-cold PBS, centrifuge at 1200 rpm for 5 min, at 2–8 °C.
2. Thaw fibrin sealant TISSEEL® sealer protein solution and thrombin solution and warm both solutions to 37 °C, as instructed by the manufacturer (*see* **Note 21**).
3. After washing the cells, gently aspirate/decant the buffer completely, without disrupting the pellet.
4. Resuspend the cells in fibrin sealant sealer protein solution based on the number of cells sorted, such that a final cell concentration of 10,000 cells/μL is achieved (the volume of sealer protein solution depends on the number of sorted cells; *see* **Note 22**).

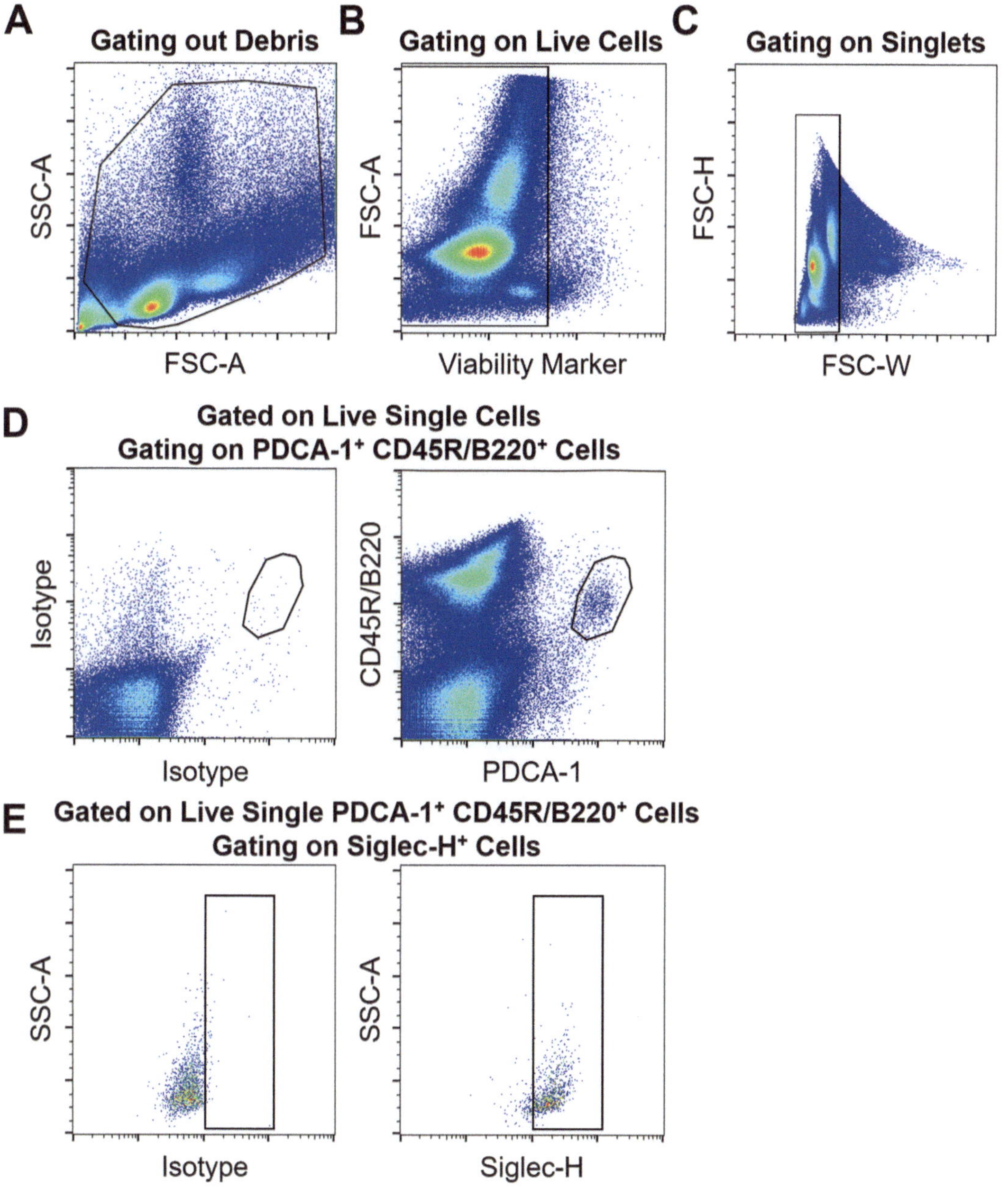

Fig. 1 Sequential gating strategy for isolation of pDCs by flow cytometry. Flow cytometric dot plots representing gating out (**a**) debris, (**b**) dead cells, (**c**) doublets, as well as (**d**) gating on PDCA-1$^+$ CD45R/B220$^+$ cells, followed by (**e**) gating on Siglec-H$^+$ cells using fluorescence minus one control (both panels in **e** are gated on live single PDCA-1$^+$ CD45R/B220$^+$ cells. In the left panel, an isotype for Siglec-H is used; however, in the right panel, an anti-Siglec-H antibody is applied). Through such sequential gating, live single PDCA-1$^+$ CD45R/B220$^+$ Siglec-H$^+$ pDCs can be sorted

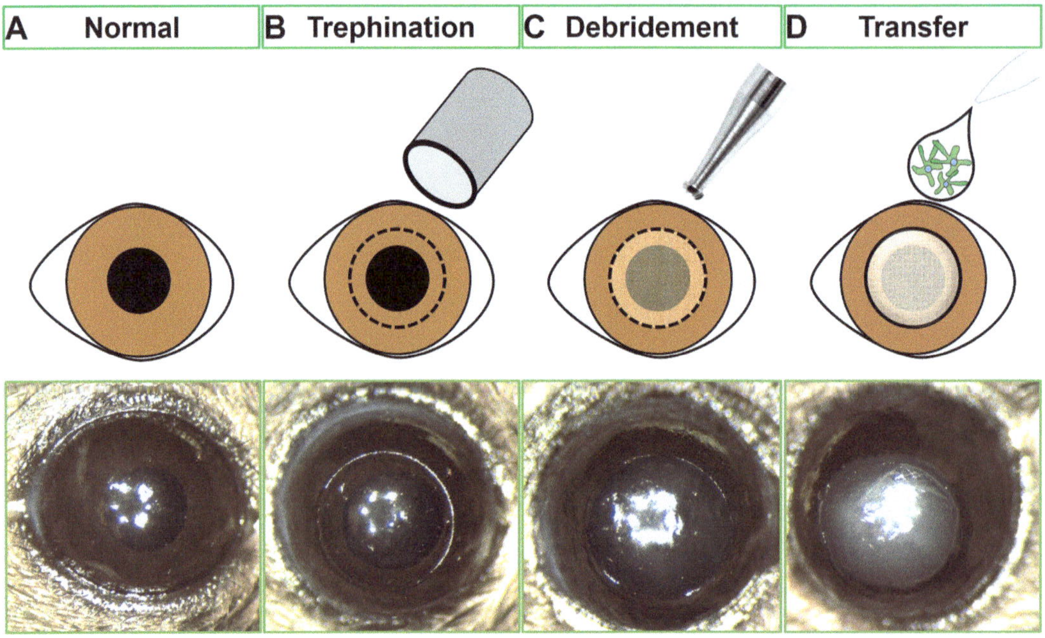

Fig. 2 Representative schematic and clinical corneal photograph illustrating the procedure. (**a**) Normal cornea prior to procedure for comparison. (**b**) Cornea following trephination using a 2 mm trephine to mark the border for epithelial debridement. (**c**) Cornea following mechanical debridement. (**d**) Typical solidified fibrin sealant on the cornea, approximately 10 min following adoptive transfer of pDCs

5. Anesthetize wild-type syngeneic (C57BL/6J) animals (*see* **Note 11**) per approved animal protocol. Confirm a deep state of anesthesia by toe pinch.
6. Apply a drop of 0.5% proparacaine ophthalmic solution on the cornea that will undergo wounding. On the contralateral cornea, apply ophthalmic artificial tears to prevent desiccation.
7. Under a dissecting microscope, stabilize the eyeball by a curved forceps and mark the central 2 mm of the cornea using a 2 mm trephine (Fig. 2; *see* **Note 23**).
8. After marking the cornea, use an Algerbrush II corneal rust ring to carefully remove the corneal epithelium (Fig. 2).
9. Apply 1 μL of the cell suspension (resuspended in sealer protein solution of fibrin sealant TISSEEL®) on the central cornea (*see* **Note 24**).
10. Quickly apply 1 μL of fibrin sealant thrombin solution (*see* **Notes 25** and **26**).
11. Wait for 5–10 min to ensure that fibrin sealant is solidified on the cornea (Fig. 2), and then apply triple antibiotic ointment on the cornea (*see* **Note 27**).
12. Different controls can be used for the adoptive transfer process. For a "sham" control, simply add 1 μL of sealer protein

solution without the cells on the wounded cornea followed by quick addition of 1 μL of thrombin solution. Irrelevant cells can be used as additional controls.

13. Place the mouse in a regular cage and provide heat by a heating lamp (or other IACUC-approved approach). Monitor the mouse until recovered from anesthesia.
14. Repeat the process for remaining animals (*see* **Notes 28** and **29**).

3.5 Outcome Assessment

1. At desired time points following corneal wounding and adoptive transfer (for instance 3, 6, 12, 24, 48, and 72 h), anesthetize the animal(s) according to the IACUC-approved protocol. Confirm a deep state of anesthesia by toe pinch.
2. Take a clinical photograph of the via slit-lamp biomicroscopy prior to fluorescein staining (optional).
3. Apply one drop of 0.25% fluorescein ophthalmic solution and leave it on the cornea for 2 min.
4. Wash the cornea using normal saline to remove excess fluorescein. Use a cotton tip applicator to avoid pouring of the excess fluorescein on the animal. Examine the cornea by slit-lamp biomicroscopy under the cobalt blue light.
5. Take photograph(s) of the cornea for visualizing the corneal fluorescein staining, hence to assess corneal epithelial defects (Fig. 3).

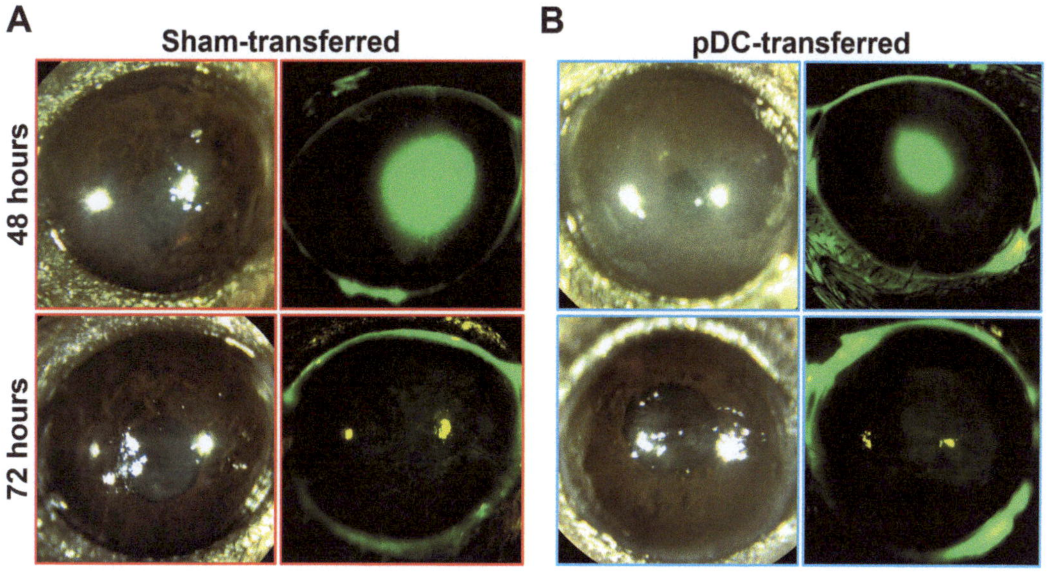

Fig. 3 Representative images of the wounded cornea. Images represent clinical corneal photograph and corneal fluorescein staining 48 and 72 h after corneal wounding and (**a**) sham transfer and (**b**) pDCs' adoptive transfer, showing smaller wound size in the cornea undergoing adoptive pDC transfer at 48 h following adoptive transfer

6. Place the mouse in a regular cage and provide heat by a heating lamp (or other IACUC-approved approach). Monitor the mouse until recovered from anesthesia.
7. Repeat the process for the remaining animals.
8. Measure the area with epithelial defect (which is stained with fluorescein) by ImageJ. In brief, open the acquired image in ImageJ. Using the "freehand selections" tool, select the stained area; on the "Analyze" tab, choose "Measure" to quantify the area with defect. Repeat similar measurements to quantify the area of the whole cornea. By dividing the defect area to the whole corneal area, the percentage of wounded area (to the total cornea) can be presented. Alternately, the corneal wound area can be measured over time and can be compared to the initial corneal wound measured at the baseline to express the percentage of wound closure (healing).

4 Notes

1. PBS free of calcium and magnesium is preferred to limit clumping, which may otherwise occur due to adhesion molecules.
2. One six-well cell culture plate can accommodate normal-sized spleens from multiple animals. However, a petri dish or a 12-well cell culture plate can also be used.
3. Although we suggest using wild-type C57BL/6J mice in this protocol, any strain of mice can be used as the source of cells. Similarly, any transgenic mice can also be used as the source of cells of interest.
4. The estimated number of donor animals needed depends on the total number of cells required for a particular experiment. From each naïve mouse spleen, the pDCs' isolation yield is expected to approximate 100,000–200,000 cells (depending on the age and gender of the mouse, technical properties of the flow cytometry sorter, and experience with the process). Our recommendation is to transfer 10,000 pDCs to each cornea. Thus, if the experiment is planned for the adoptive transfer of pDCs to five mice, $5 \times 10{,}000 = 50{,}000$ pDCs are needed. However, if more than 10 mice are planned for adoptive transfer of pDCs, more donor mice may be needed. Alternatively, prior treatment of donor mice with FLT3 ligand (for instance 10 μg subcutaneously) daily for 7–10 days [26] or administration of tumor cells expressing the FLT3 ligand to the donor approximately 10–14 days prior to euthanizing the animal can enhance pDC density and thus isolation yield [27].
5. It is recommended to prepare ACK RBC lysis buffer in advance and store at 2–8 °C for up to 3–6 months.

6. It is recommended to prepare FACS buffer in advance and store at 2–8 °C for up to 3–6 months. Optionally, DNase I (25–50 μg/mL) can be added to the FACS buffer to further avoid cell clumping in the process of single cell preparation.
7. A variety of viability markers are available, such as 7-aminoactinomycin D (7-AAD), propidium iodide (PI), and amino-labeling viability dyes. For selecting one reagent, factors such as prior experience, technical specifications of flow cytometry sorter, potential spectral overlaps with fluorophore-conjugated antibodies, which may sophisticate the isolation process, and costs may be considered. Also, considering that sorted cells will be used for in vivo applications, it is vital to use nontoxic viability dyes dissolved in nontoxic solvents to minimize the effect of the marker on the viability and function of sorted cells.
8. In this protocol, we describe adoptive transfer of pDCs, a relatively rare population of immune cell type; however, the technique can be expanded to adoptive transfer of other cell types isolated by flow cytometry, commercially available isolation kits, or ex vivo generated or manipulated cells of interest.
9. Since murine pDCs harbor a unique cell surface repertoire, application of multiple cell markers is necessary to validly distinguish them; in this protocol, we present the minimum combination of markers to isolate pDCs (PDCA-1, CD45R/B220, and Siglec-H), to minimize the potential effect of the antibodies on the pDC function. However, if sources other than bone marrow, spleen, and lymph nodes are used to isolate pDCs, addition of anti-CD45 antibody, as a pan-leukocyte marker, is recommended. Further, addition of other markers such as CD11b, F4/80, CD3, and CD19 (in which pDCs lack) as well as CD11c, Ly6C, and Ly49Q (expressed by pDCs) might be necessary to differentiate pDCs.
10. Given that the various available cell sorting flow cytometers have different technical configurations and laser settings, we leave the choice of fluorophores and panel design to the readers. Importantly, panels should be designed carefully according to standard recommendations. In this regard, application of appropriate isotype controls, single stained controls, and fluorescence minus one controls are necessary for proper isolation of cells. In Fig. 1, we used PE-conjugated anti-PDCA-1, PE-Cy7-conjugated anti-CD45R/B220, and FITC-conjugated anti-Siglec-H.
11. Similar to **Note 3**, although in this protocol we suggest using wild-type C57BL/6J mice, any transgenic mouse and any strain of mice can be used, as long as syngeneic donors are

used as the source of cells of interest (to avoid potential allogeneic immune response).

12. Other equivalent reagents can be used, such as fluorescein sodium strips, which can be placed in appropriate amounts of normal saline to yield 0.25% fluorescein solution.
13. If euthanizing multiple animals, multiple wells may be required. Each well of a six-well cell culture plate can fit 4–5 normal mouse spleens.
14. In order to decrease the duration of the procedure, it is recommended to continue with the first step of the next process (Subheading 3.3) while the sample is being centrifuged.
15. This second wash step is recommended to eliminate any clumping/debris that may arise from the preceding lysis step.
16. Blocking with FACS buffer containing 1:100 anti-CD16/CD32 Fc block is crucial to prevent nonspecific binding of fluorophore-conjugated antibodies (used in the next steps) to Fc receptors expressed on various immune cells.
17. In order to decrease the duration of the procedure, it is recommended to continue with preparation of the next steps (**steps 3** and **4**; labeling the tubes, taking the antibodies from the refrigerator, and arranging them on an ice-cold container) while the samples are in incubation.
18. This filtering step is recommended to remove any potential cell aggregates/clumps formed during the prior steps to prevent clogging the flow cytometry sorter.
19. Although only 10,000 pDCs are required for adoptive transfer to each mouse, considering the potential loss of cells during the following washing steps and potential loss of cells during the adoptive transfer by inexperienced users, we recommend to sort at least 150,000–200,000 pDCs, even if only one mouse is planned to receive the cells.
20. In order to decrease the duration of the procedure, it is recommended to continue with preparation of **step 2** of the next section (Subheading 3.4; thawing fibrin sealant) while the pDCs are being sorted.
21. As mentioned by the manufacturer, this heating step to 37 °C is important for optimal enzymatic activity of the sealer protein solution and the thrombin solution.
22. For instance, if 200,000 pDCs are sorted, add 20 μL of fibrin sealant sealer protein solution. As another example, if 250,000 pDCs are sorted, add 25 μL of fibrin sealant sealer protein solution.
23. Trephines with a smaller diameter (for instance 1.5 mm) can also be used to mark the cornea and subsequently to create a

smaller wound by debriding the epithelium. However, it should be considered that the smaller the wound, the faster it heals. Thus, if exploring the effect of another therapy (for instance, adoptive transfer of another cell type), which potentially has a small effect size, the power of the technique in detecting the differences between control and treatment groups will be limited. Also, it should be noted that if the aim of adoptive transfer is studying the role of cells of interest in pathogenesis or treatment of a corneal condition other than wound healing, generally, a small area of epithelial debridement is sufficient to allow transfer of cells.

24. Upon transferring the cells in the sealer protein solution of fibrin sealant to the cornea, the solution may slip from the central cornea. Ensuring proper depth of anesthesia, stabilizing the cornea with a curved forceps, or preventing potential hand shaking by supporting the pipet tip with the opposite hand can improve the process and success of the procedure.
25. When sealer protein solution and thrombin solution are mixed, the components will react to form a fibrin seal, securing the cells within the epithelial defect (Fig. 2).
26. Similar to sealer protein solution, the applied thrombin solution may slip from the cornea. As mentioned in **Note 24**, ensuring proper depth of anesthesia, stabilizing the cornea with a curved forceps, or preventing potential hand shaking by supporting the pipet tip with the opposite hand can improve the process and success of the procedure.
27. If more than one mouse is planned for adoptive transfer of pDCs, the procedure can be started on the next animal while the fibrin sealant is solidifying on the first mouse to accelerate the experiment.
28. It should be considered that the process of donor tissue collection, tissue processing and single cell preparation, cell staining, and sorting pDCs may take about 5 h. Thus, for performing corneal wounding and adoptive transfer of pDCs in multiple animals, it should be considered that the longer the process takes (the more the number of mice under the experiment), the more pDCs may undergo apoptosis, reducing the efficacy of the adoptive transfer of pDCs for corneal wound healing.
29. To assess if adoptive transfer of pDCs is successful, cells can be stained using a commercially available cell tracer prior to adoptive transfer and the corneas can be assessed at the desired time points to measure the density of the transferred cells and their longevity in the cornea. Nevertheless, considering that many of the commercially available cell tracers are toxic or are dissolved in toxic solvents, if pDCs are stained with such tracers, we do not recommend that cell function be studied after adoptive

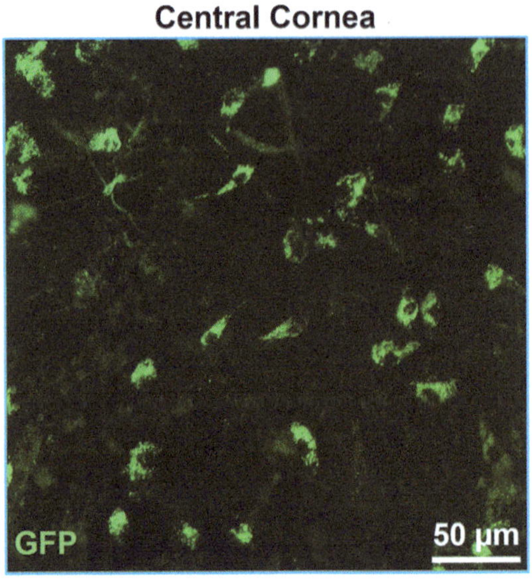

Fig. 4 Integration of adoptively transferred pDCs to corneal stroma. A DPE-GFP × RAG1$^{-/-}$ mouse with specifically GFP-tagged pDCs was used as the source of pDCs in this experiment. A representative corneal whole-mount micrograph, 48 h following adoptive transfer shows integration of GFP-tagged adoptively transferred pDCs to the recipient cornea

transfer, should tracers be used for verification of successful transfer. Alternatively, transgenic mice with fluorescently tagged pDCs such as DPE-GFP × RAG1$^{-/-}$ mice, in which pDCs are tagged with GFP [28], can be used for isolation of pDCs and tracking cells after adoptive transfer (Fig. 4).

Acknowledgments

The authors would like to appreciate the technical assistance of Allen Parmelee and Stephen Kwok at Tufts University Flow Cytometry Core for their assistance in sorting the cells. This work was supported by the following grants to P.H.: NIH R01-EY022695, NIH R01-EY026963, NIH R01-EY029602, Massachusetts Lions Eye Research Fund, and Tufts Medical Center Institutional Support.

References

1. Swiecki M, Colonna M (2010) Unraveling the functions of plasmacytoid dendritic cells during viral infections, autoimmunity, and tolerance. Immunol Rev 234(1):142–162. https://doi.org/10.1111/j.0105-2896.2009.00881.x
2. Won HY, Lee JY, Ryu D et al (2019) The role of plasmacytoid dendritic cells in gut health. Immune Network 19(1):e6. https://doi.org/10.4110/in.2019.19.e6
3. Rogers NM, Isenberg JS, Thomson AW (2013) Plasmacytoid dendritic cells: no longer an enigma and now key to transplant tolerance? Am J Transplant 13(5):1125–1133. https://doi.org/10.1111/ajt.12229

4. Li S, Wu J, Zhu S et al (2017) Disease-associated plasmacytoid dendritic cells. Front Immunol 8:1268. https://doi.org/10.3389/fimmu.2017.01268
5. Sendra VG, Jamali A, Harris DL et al (2014) Role of plasmacytoid dendritic cell in the immune regulation in sutured inflamed cornea. Invest Ophthalmol Vis Sci 55(13):1694–1694
6. Sendra V, Jamali A, Harris DL et al (2015) Plasmacytoid dendritic cells mediate adaptive immunity in acute herpes simplex virus keratitis. Invest Ophthalmol Vis Sci 56 (7):1856–1856
7. Sendra VG, Jamali A, Lopez MJ et al (2017) Plasmacytoid dendritic cells modulate corneal inflammation through transforming growth factor (TGF)-β1. Invest Ophthalmol Vis Sci 58(8):3618–3618
8. Jamali A, Lopez MJ, Sendra V et al (2016) Plasmacytoid dendritic cells maintain corneal heme-angiogenic privilege through secretion of anti-angiogenic molecules. Invest Ophthalmol Vis Sci 57(12):1430–1430
9. Jamali A, Lopez MJ, Sendra V et al (2015) Plasmacytoid dendritic cells demonstrate vital neuro-protective properties in the cornea and induce corneal nerve regeneration. Invest Ophthalmol Vis Sci 56(7):4355–4355
10. Jamali A, Lopez MJ, Sendra VG et al (2017) Local adoptive transfer of plasmacytoid dendritic cells as a novel therapeutic approach for corneal nerve regeneration. Invest Ophthalmol Vis Sci 58(8):993–993
11. Verhagen C, Mor F, Kipp JB et al (1999) Experimental autoimmune keratitis induced in rats by anti-cornea T-cell lines. Invest Ophthalmol Vis Sci 40(10):2191–2198
12. He Y, Jie Y, Wang B et al (2010) Adoptive transfer of donor corneal antigen-specific regulatory T cells can prolong mice corneal grafts survival. Cornea 29(Suppl 1):S25–S31. https://doi.org/10.1097/ICO.0b013e3181ea4999
13. Fitzpatrick S, Lausch R, Barrington RA (2019) CCR6-positive gammadelta T cells provide protection against intracorneal HSV-1 infection. Invest Ophthalmol Vis Sci 60 (12):3952–3962. https://doi.org/10.1167/iovs.19-27810
14. Sarkar R, Mathew A, Sehrawat S (2019) Myeloid-derived suppressor cells confer infectious tolerance to dampen virus-induced tissue immunoinflammation. J Immunol 203 (5):1325–1337. https://doi.org/10.4049/jimmunol.1900142
15. Chauhan SK, Saban DR, Dohlman TH et al (2014) CCL-21 conditioned regulatory T cells induce allotolerance through enhanced homing to lymphoid tissue. J Immunol 192 (2):817–823. https://doi.org/10.4049/jimmunol.1203469
16. Guo X, Jie Y, Ren D et al (2012) In vitro-expanded CD4(+)CD25(high)Foxp3(+) regulatory T cells controls corneal allograft rejection. Hum Immunol 73(11):1061–1067. https://doi.org/10.1016/j.humimm.2012.08.014
17. Martinez-Carrasco R, Sanchez-Abarca LI, Nieto-Gomez C et al (2019) Subconjunctival injection of mesenchymal stromal cells protects the cornea in an experimental model of GVHD. Ocular Surface 17(2):285–294. https://doi.org/10.1016/j.jtos.2019.01.001
18. Shao C, Chen Y, Nakao T et al (2019) Local delivery of regulatory T cells promotes corneal allograft survival. Transplantation 103 (1):182–190. https://doi.org/10.1097/tp.0000000000002442
19. Nakao S, Hafezi-Moghadam A, Ishibashi T (2012) Lymphatics and lymphangiogenesis in the eye. J Ophthalmol 2012:783163. https://doi.org/10.1155/2012/783163
20. Hendricks RL, Epstein RJ, Viana MA et al (1991) A reproducible method for injecting the mouse corneal stroma. Invest Ophthalmol Vis Sci 32(2):366–370
21. Epstein RJ, Stulting RD (1987) Corneal neovascularization induced by stimulated lymphocytes in inbred mice. Invest Ophthalmol Vis Sci 28(9):1505–1513
22. Matthaei M, Meng H, Bhutto I et al (2012) Systematic assessment of microneedle injection into the mouse cornea. Eur J Med Res 17:19. https://doi.org/10.1186/2047-783x-17-19
23. Jamali A, Lopez MJ, Harris DL et al (2019) Local adoptive transfer of plasmacytoid dendritic cells as a novel therapeutic approach for corneal neovascularization. Invest Ophthalmol Vis Sci 60(9):897–897
24. Griffith GL, Kasus-Jacobi A, Lerner MR et al (2014) Corneal wound healing, a newly identified function of CAP37, is mediated by protein kinase C delta (PKCdelta). Invest Ophthalmol Vis Sci 55(8):4886–4895. https://doi.org/10.1167/iovs.14-14461
25. Li Z, Burns AR, Smith CW (2006) Two waves of neutrophil emigration in response to corneal epithelial abrasion: distinct adhesion molecule requirements. Invest Ophthalmol Vis Sci 47 (5):1947–1955. https://doi.org/10.1167/iovs.05-1193
26. Karsunky H, Merad M, Cozzio A et al (2003) Flt3 ligand regulates dendritic cell development from Flt3+ lymphoid and myeloid-

committed progenitors to Flt3+ dendritic cells in vivo. J Exp Med 198(2):305–313. https://doi.org/10.1084/jem.20030323

27. Mach N, Gillessen S, Wilson SB et al (2000) Differences in dendritic cells stimulated in vivo by tumors engineered to secrete granulocyte-macrophage colony-stimulating factor or Flt3-ligand. Cancer Res 60(12):3239–3246

28. Iparraguirre A, Tobias JW, Hensley SE et al (2008) Two distinct activation states of plasmacytoid dendritic cells induced by influenza virus and CpG 1826 oligonucleotide. J Leukoc Biol 83(3):610–620. https://doi.org/10.1189/jlb.0807511

Chapter 17

Murine Corneal Epithelial Wound Modeling

Dhara Shah and Vinay Kumar Aakalu

Abstract

A murine model of corneal epithelial wounding can be performed using simple injury and imaging methods. Here, we describe the creation of a central corneal epithelial defect using mechanical debridement under ophthalmic microscopic visualization. Subsequent monitoring with vital dye application and slit-lamp bio microscopy (slit-lamp) is described in detail.

Key words Cornea, Wound healing, Epithelial, Murine, Mouse, Injury, Ocular surface

1 Introduction

Corneal epithelial wounds can be modeled in mice using standard wounding methods (Fig. 1). These injuries are simple, repeatable, easily imaged, and are useful for drug development and pathophysiologic investigation. Multiple methods can be used to create a full-thickness epithelial (partial thickness of the cornea) wound including mechanical, chemical, or LASER debridement, depending on the goals of the experiment [1–3]. Wounding of deeper layers of the cornea is also utilized in studies of stromal healing, scarring responses, and other aspects of corneal biology [4]. These injuries are easily performed under ophthalmic microscopic guidance and imaged using standard slit-lamp imaging and vital dyes (Figs. 1 and 2). The fast rate of closure and generally good reproducibility of these injuries allow for rigorous testing of interventions to promote or delay wound healing or testing of the physiology of wound healing [1–3].

Here, we describe a mechanical method of debridement and imaging-based follow-up to document injury responses and wound-healing rates over time.

Hiranmoy Das (ed.), *Wound Regeneration: Methods and Protocols*, Methods in Molecular Biology, vol. 2193,
https://doi.org/10.1007/978-1-0716-0845-6_17,

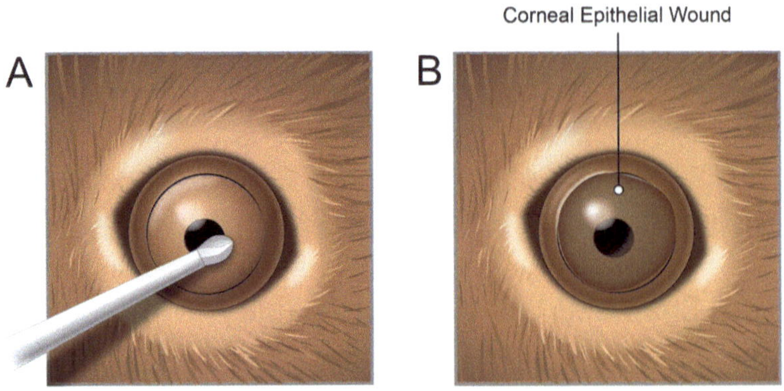

Fig. 1 Mouse model of corneal epithelial injury. (**a**) Mouse eye depicted with intact corneal epithelium and Algerbrush® II. Black circle indicates demarcated region of planned wound using a 2.0-mm disposable biopsy punch. (**b**) Central corneal epithelial wound created by Algerbrush® II debridement

2 Materials

1. Six- to twenty-week-old C57BL6 wild-type mice (Jackson Laboratory, Bar Harbor, ME; Charles River, Wilmington, MA). *Alternative mouse strains can be chosen based on specific experimental needs.
2. 27 G, 1/2″ sterile needles and 1-ml sterile syringes for intraperitoneal (IP) injections.
3. Weight scale.
4. Ketamine and xylazine (weight-based dosing as per veterinarian recommendation).
5. Buprenorphine HCL injection (weight-based dosing as per veterinarian recommendation).
6. 0.5% Proparacaine hydrochloride sterile.
7. Betadine solution (diluted as per local protocol).
8. Ophthalmic antibiotic ointment.
9. Saline solution.
10. 3 cm^3 syringe.
11. 4″ × 4″ sterile gauze.
12. Surgical scissors.
13. Hot bead sterilizer (Germinator 500, Electron microscopy Sciences, Hatfield, PA).
14. Heated pad.
15. Merocel eye spear (Beaver Visitec Intl Ltd., Waltham, MA).
16. Uni-Punch Seamless Disposable Biopsy punch (Premier, Plymouth Meeting, PA).

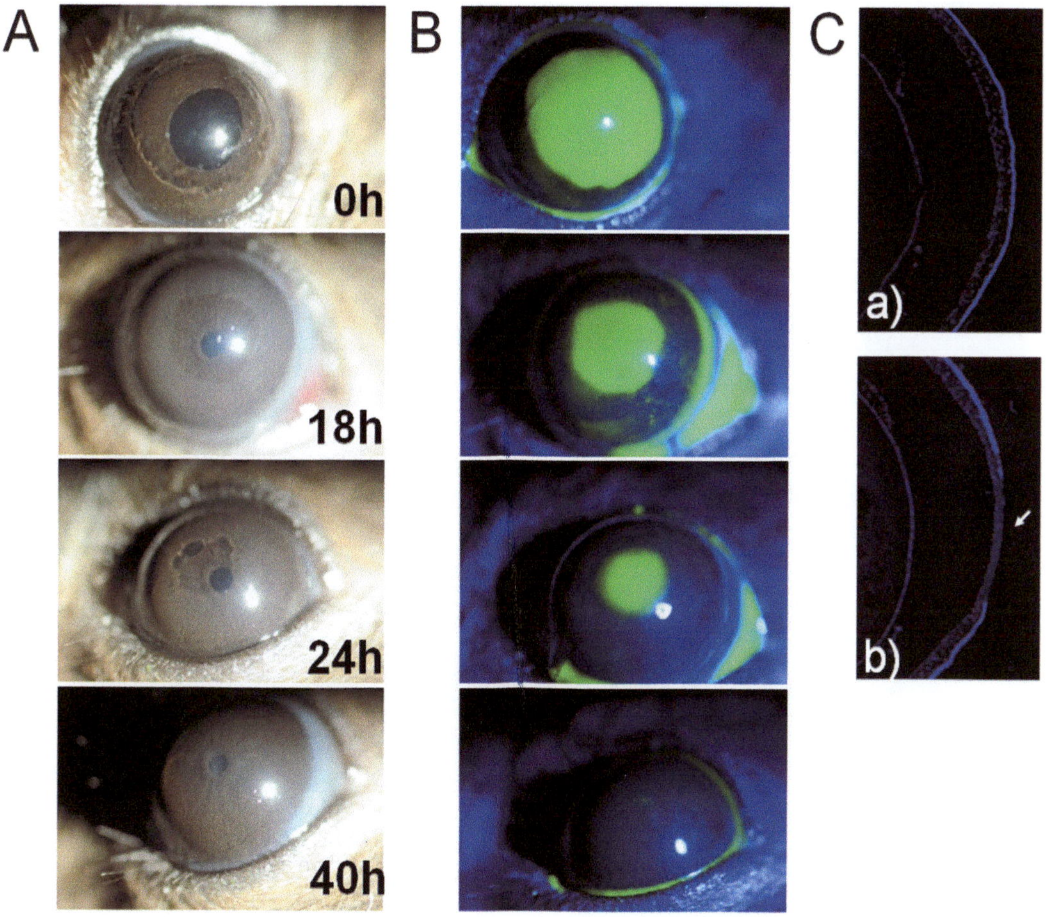

Fig. 2 Slit-lamp images (**a**) Wide beam slit-lamp image of epithelial wound in mouse cornea, and progressive healing over 40 hours (top to bottom). (**b**) Cobalt light illuminated slit-lamp images of corneal epithelial wound, as highlighted by fluorescein application. (**c**) Cross-sections of anterior segment of murine eye, showing intact epithelium (**a**) and corneal epithelial defect (**b**) shown with arrow

17. Algerbrush® II (The Alger Company, Lago Vista, TX) (Fig. 3a).
18. Fluorescein (FUL-GLO Fluorescein Sodium ophthalmic strips, Akorn, Lake forest, IL) diluted in basic salt solution (BSS).
19. Ophthalmic surgical microscope (Fig. 3b).
20. Slit-lamp biomicroscope (Nikon FS-2 photo slit-lamp with a Nikon D200 camera) (Melville, NY) (Fig. 4).
21. Image J software (Image J 1.47v, NIH, Thornwood, Bethesda, MD).
22. Personal protective equipment: hair net, face mask, gloves, surgical gown and shoe covers.

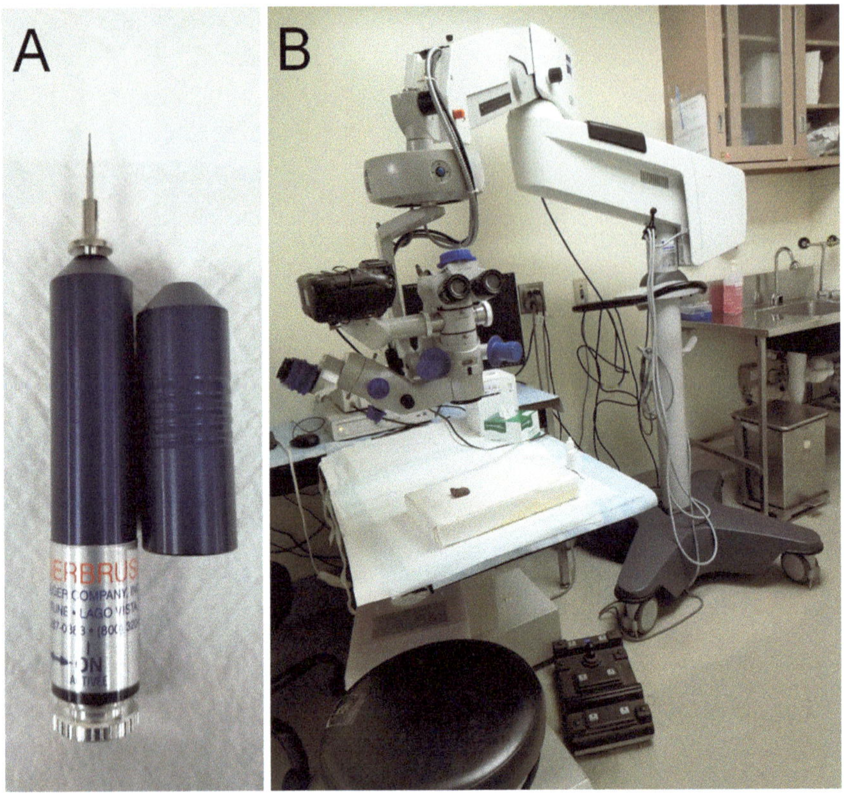

Fig. 3 Algerbrush® II and surgical microscope. (**a**) Algerbrush® II used for corneal epithelial wounding. (**b**) Ophthalmic stereoscopic surgical microscope with surgical bed and standard setup

3 Methods

3.1 Anesthesia

1. The mice will be placed under anesthesia by IP injection of ketamine and xylazine, as approved by local vivarium/protocol and by weight.
2. The state of anesthesia will be monitored by the lack of response to toe pinch and lack of blinking in response to the placement of eye drops. If the animal is found to react, then additional anesthesia may need to be given, per local protocol.
3. Apply one drop of proparacaine hydrochloride (0.5%) to the experimental eye multiple times over a few minutes. Give analgesic immediately after induction of anesthesia.

3.2 Surgical Technique

1. Antiseptic Preparation of the Surgical Site: Apply diluted betadine solution, topically using a 3-cm^3 syringe to the ocular surface and fornices three times.
2. Prepare periocular skin/fur with the same solution using 4″ × 4″ sterile gauze and diluted betadine solution.

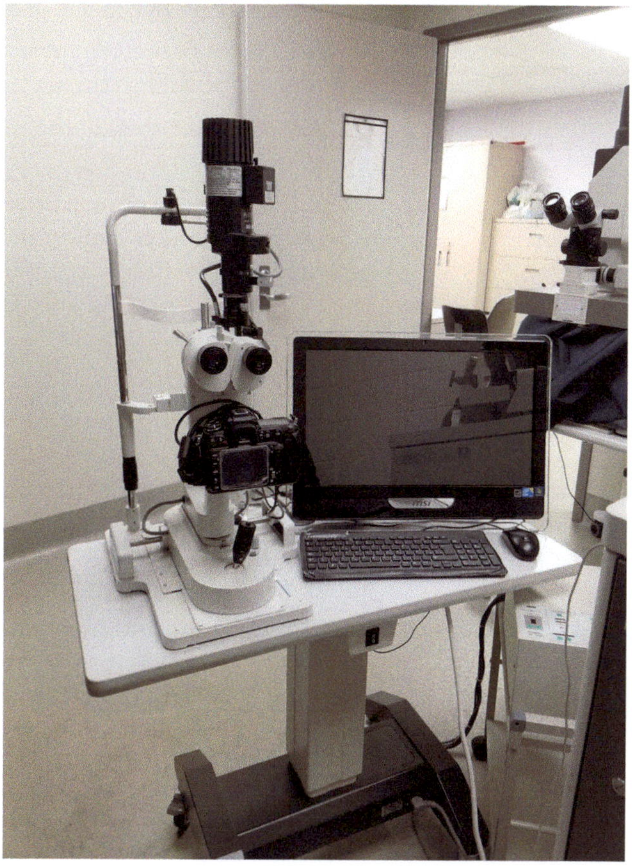

Fig. 4 Slit-lamp photography setup with monitor for viewing images

3. Use a surgical microscope, and surgical scissors cut any long eyelashes that may interfere with visualization.
4. Use hot bead sterilizer in between mice to sterilize the scissors; take care to avoid touching mice with hot instruments.
5. Then, use a sterile 2.0-mm disposable biopsy punch to gently depress and demarcate the central epithelium, without penetrating deeper than the epithelial layer.
6. Remove the demarcated region of the central epithelium using an Algerbrush® II, taking care not to penetrate deeper than Bowman's layer.

3.3 Post-operative Care

1. Immediately after the procedure, perform image acquisition using slit-lamp imaging (depending on the protocol, preoperative visualization and image capture may be necessary).
2. Using a wide beam, take enface images of the eye at a standardized distance from the slit-lamp (to ensure close comparability of images between animals and time points in the experiment).

3. Apply fluorescein dye as a 5-μl eye drop to the injured eye.
4. Remove excess fluorescein by using a Merocel eye spear, taking care to avoid touching the surface of the eye.
5. Use a cobalt filter to take the image of the fluorescein-stained injury area.
6. Ophthalmic ointment may be placed to help promote healing, or experimental intervention material, therapeutics, etc. could be applied at this time, depending experimental protocol.
7. Put the animal on the heating pad until awake.
8. Once mice regain consciousness, give analgesic (e.g., Buprenorphine HCL injection). Also give this pain reliever on the following day postsurgery, or as recommended by the local vivarium.

3.4 Monitoring and Measurement

1. All mice should be monitored at least once daily by the veterinary or lab personnel after surgery for discomfort, distress as noted below until the end of the study.
2. Monitor daily for any signs of humane endpoint criteria, for example, infection (crusting, redness, hazy cornea, purulent discharge, etc.), discomfort or distress such as lethargy, ruffled fur, huddling, abnormal behavior, body posture, guarding, rubbing of eyes, excessive tearing, and decreased food or water intake.

4 Notes

1. Make sure to work with the local vivarium, veterinarian, and regulatory agencies to determine the safest protocol for anesthesia, operative and postoperative care.
2. Some anesthetics are controlled substances, depending on the country and local regulations, and should be maintained in a locked or limited-access location.
3. Weight-based anesthetic delivery should be followed and contemporaneous/daily weights should be performed to ensure that the correct dose of medication is given.
4. Sterile equipment, freshly prepared and packaged medications, and sterile field preparation and maintenance are critically important to avoid animal infection or unintended consequences.
5. Make sure to wear all appropriate personal protective equipment and take care to avoid sharps injuries.
6. Imaging with a slit-lamp multiple times over a short time course may pose a challenge with regard to repeated anesthesia.

Local veterinarian consultation should be undertaken to determine the best course of action. A gentle, trained hand can be used to restrain mice for imaging at the slit-lamp without anesthesia but should be done with care and after training.

7. Frequency of imaging should depend on the predicted time course of the experiment. Larger wounds or wounds in healing-deficient mice or inflamed corneas may take longer. Typically, in healthy wild-type mice (C57BL6), a 2-mm corneal wound will heal by 30 h or sooner [5].
8. Care should be taken to try and standardize the distance of mice from the slit-lamp and maintain brightness settings and magnification throughout the experiment to allow for easy comparison throughout or between experiments.
9. Monitoring for human endpoint criteria or other unanticipated outcomes in survival surgeries is critical and should be preplanned carefully with local authorities.
10. Euthanasia should be planned and clear protocols developed with institutional officials.
11. It should be noted that many other methods exist for creating corneal epithelial wounds, some of which are appropriate for chemical injuries, injuries which promote corneal neovascularization or are otherwise specific to different experimental needs [1–3].

Acknowledgments

Illustrations courtesy of Lauren Kalinoski MA, University of Illinois, Department of Ophthalmology and Visual Sciences.

References

1. Putra I, Rabiee B, Anwar KN et al (2019) *Staphylococcus aureus* alpha-hemolysin impairs corneal epithelial wound healing and promotes intracellular bacterial invasion. Exp Eye Res 181:263–270
2. An S, Raju I, Surenkhuu B et al (2019) Neutrophil extracellular traps (NETs) contribute to pathological changes of ocular graft-vs.-host disease (oGVHD) dry eye: Implications for novel biomarkers and therapeutic strategies. Ocul Surf 17(3):589–614
3. Carlson E, Kao WW, Ogundele A (2018) Impact of hyaluronic acid-containing artificial tear products on reepithelialization in an in vivo corneal wound model. J Ocul Pharmacol Ther 34 (4):360–364
4. Ljubimov AV, Saghizadeh M (2015) Progress in corneal wound healing. Prog Retin Eye Res 49:17–45
5. Shah D, Son KN, Kalmodia S, Lee BS, Ali M, Balasubramaniam A, Shukla D, Aakalu VK (2020) Wound healing properties of histatin-5 and identification of a functional domain required for histatin-5 induced cell migration. Molecular Therapy-Methods & Clinical Development.

Chapter 18

Infection-Induced Porcine Ex Vivo Corneal Wound Model to Study the Efficacy of Herpes Simplex Virus-1 Entry and Replication Inhibitors

Tejabhiram Yadavalli, Raghuram Koganti, and Deepak Shukla

Abstract

Corneal infections by viruses and bacteria can result in ocular surface defects, ulcers, or wounds. Herpes simplex virus type-1 (HSV-1) is a human virus with global seroprevalence in the range of 60–90%. While the virus more commonly causes mucocutaneous lesions including ulcers on the face and mouth, it is also a leading cause of infection-associated blindness. In this chapter, we discuss an in-depth protocol required to evaluate corneal damage due to HSV-1 infection using porcine models of ex vivo infection. Our methods can be adapted to study similar infections caused by other viruses and bacteria.

Key words HSV-1, Ex vivo, Cornea, Porcine model

1 Introduction

Ocular infections are a common cause for vision loss [1]. HSV has a globally high seroprevalence and many of those infected individuals suffer from recurrent eye infections, resulting in corneal scarring, neovascularization, and stromal keratitis [2–6]. HSV infections might also lead to other serious but less frequent diseases such as retinitis, encephalitis, and systemic morbidity among untreated immunocompromised patients [7–9]. While acyclovir (ACV) and its nucleoside analogs are successful in curbing HSV-1 infection and control reactivation, continuous use of these drugs causes undesired side effects including renal failure, fetal abnormalities, and fails to prevent emergence of drug-resistant viral strains in the long run [10–14]. In order to study the complexity of HSV-1 ocular infection, investigators typically employ in vitro models such as monolayers and 3D cultures of human corneal cells or in vivo models such as murine or rabbit ocular infections. While in vitro models provide significant input regarding the cell biology of HSV-1 infections, they lack the intricacy of a corneal tissue and

Hiranmoy Das (ed.), *Wound Regeneration: Methods and Protocols*, Methods in Molecular Biology, vol. 2193, https://doi.org/10.1007/978-1-0716-0845-6_18,

may not shed light on dendritic spread of the virus [15]. In vivo systems, on the other hand, provide much more insight into corneal spread and immune activation during HSV-1 infection; however, the model needs trained investigators and large animal care facilities to overlook the experiments being conducted. In this chapter, we discuss the use of an ex vivo model to study HSV-1 infection-induced wound system using porcine corneas. These tissue explant cultures can be used not only to study the cell biology and molecular biology of an infection-induced wound system but also to study the pharmacology of potential drugs to treat these herpetic wounds [18]. This system can easily be adapted for use with other viral and bacterial infections as well.

2 Materials

2.1 Cell Culture

1. Vero cell line to grow the virus.
2. HSV-1 (KOS) 17 green fluorescence protein (GFP)-expressing version of KOS strain with GFP on a cytomegalovirus (CMV) promoter. This virus can infect, replicate, and egress from one cell to another, producing GFP in every cell it infects. This virus can be used to visualize epithelial dendrite formation in the cornea.
3. Porcine corneas: Freshly sacrificed porcine corneas were obtained from an approved vendor on the same day the experiment was conducted.
4. Phosphate-buffered saline (PBS).
5. Dulbecco's modified Eagle's medium (DMEM).
6. Minimum Essential Medium (MEM).
7. OptiMEM.
8. Fetal bovine serum (FBS).
9. Penicillin/streptomycin.
10. Insulin–Transferrin–Selenium.
11. Antimycotic and antibiotic (AA). This solution contains 10,000 units/mL of penicillin, 10,000 μg/mL of streptomycin, and 25 μg/mL of amphotericin B.

2.2 Reagents

1. Ultra-pure phenol chloroform isoamyl alcohol mix.
2. Proteinase-K.
3. Molecular biology grade ethanol.
4. Molecular biology grade 2-propanol.
5. Molecular biology grade chloroform.
6. Trizol.

7. Molecular biology grade water.
8. High-capacity cDNA reverse transcription kit.
9. *N′*,*N′*-dimethyl-*N*-[4-[(*E*)-(3-methyl-1,3-benzothiazol-2-ylidene)methyl]-1-phenylquinolin-1-ium-2-yl]-N-propylpropane-1,3-diamine (SYBR) green master mix.
10. Methylcellulose.
11. Crystal violet tablets.
12. Methanol.

2.3 Polymerase Chain Reaction (PCR) Primers

PCR primers were purchased from Integrated DNA Technologies, Inc. (IDT), USA. The PCR primers were diluted to a 100-μM stock concentration and stored at −20 °C for long-term storage (*see* **Note 1**). Working solution of the primers were prepared at 2.5 μM concentration using molecular biology grade water and stored up to a month at −20 °C [15].

2.4 Instrumentation

1. Freezers −20 and − 80 °C.
2. Refrigerator 4 °C.
3. Heating block capable of heating to 90 °C.
4. Biosafety cabinet with a biosafety level-2 (BSL-2) certification.
5. Autoclave.
6. Weighing balance 1–100 g.
7. Stirring magnet.
8. Magnetic stirrer.
9. 500-mL glass bottle.
10. Micropipettes 1–1000 μL.
11. Cell scraper—25 cm.
12. Puritan Calgiswab 6″ sterile calcium alginate standard swabs—20.
13. Metallic scissors.
14. Serological pipetting equipment.
15. Serological pipettes, 5–25 mL.
16. Tissue culture flasks, T175 cm^2.
17. Tissue culture treated plates (6-well).
18. Tissue culture incubators which can maintain 5% CO_2 and 37 °C temperature.
19. Fresh box of blades.
20. 30-g needles.
21. Flat edged tweezers—2.
22. Procedure bench covers—as needed.

23. 10× inverted light microscope to visualize cells.
24. Probe sonicator (500 W, 20 kHz).
25. Benchtop vortexer.
26. 384-well plates and plate covers.
27. NanoDrop nucleic acid quantification system.
28. qPCR machine—QuantStudio 7.
29. Tabletop micro-centrifuge—high speed 15,000 × *g*.
30. Centrifuge that can fit 10- and 50-mL tubes (up to 4000 × *g*).
31. Color camera to take images of the corneas. SteREO Discovery V20 microscope.

2.5 Media for Growing Vero Cells–DMEM Whole Media

1. Inside the tissue culture hood, open DMEM media bottle (*see* **Note 2**). Using a serological pipette, remove 55 mL of media from the bottle and discard. Add 50 mL of FBS and 5 mL of P/S to DMEM. Refrigerate at 4 °C.

2.6 Media for Plaque Assay—5% Methylcellulose Media

1. Autoclave 2.5 g of methylcellulose with a stirring magnet in a 500-mL glass bottle. Let the bottle cool down at room temperature (*see* **Note 3**). Inside the tissue culture hood, open a DMEM media bottle. Using a serological pipette, remove 55 mL of media from the bottle and discard. Add 50 mL of FBS and 5 mL of P/S to DMEM. Add 500 mL of this media to the glass bottle that contains methylcellulose. Stir at 4 °C for 2 days using a magnetic stirrer. Refrigerate at 4 °C (*see* **Note 4**).

2.7 Media for Porcine Cornea

1. Inside the tissue culture hood, open the media bottle (MEM). Remove 10 mL of MEM media and discard using a serological pipette. Add 5 mL of ITS and 5 mL of AA to the media. Refrigerate at 4 °C.

2.8 Crystal Violet Stock and Working Solution

1. To make a stock solution, add 1 g of crystal violet powder to 100 mL of 20% ethanol in water (*see* **Notes 5** and **6**). The stock solution can be stored in dark for greater than a year. To make a working solution from the stock, add 50 mL of the stock solution to 350 mL of water. Add 100 mL of ethanol to the mix in point 3 to make a 500 mL working crystal violet solution. Store both stock and working solutions in the dark at room temperature.

3 Methods

3.1 Growing the HSV-1 17-GFP Virus

1. Grow a confluent layer of Vero cells in a T175 flask, approximating 20 million cells.

2. Add approximately two million 17 GFP infectious virus particles to 10 mL of OptiMEM.
3. Wash the monolayer of cells in the T175 flask with 10 mL PBS.
4. Add the infected OptiMEM to the monolayer of 79VB4 cells.
5. Incubate the infected cells in the tissue culture incubator for 2 h.
6. Aspirate the OptiMEM and wash the monolayer with 10 mL PBS.
7. Add 20 mL DMEM to the infected monolayer of cells.
8. Incubate for 72 h or until the cells look rounded.
9. Using a cell scraper, carefully scrape all the cells from the flask.
10. Collect the cells in to a 50-mL centrifuge tube.
11. Centrifuge the cells at 800 × *g* for 30 min.
12. Using a vacuum aspirator or serological pipette, carefully remove most of the supernatant.
13. In 0.5 mL of fresh DMEM media, resuspend the pellet.
14. Sonicate the cell pellet at 30% amplitude with 5-s pulses six times with a gap of 3 s between pulses (*see* **Note** 7).
15. Place on ice immediately.
16. For short-term storage of the virus, store at 4 °C up to 3 days.
17. For long-term storage, add equal volume of autoclaved milk prior to placing in −80 °C (*see* **Note 8**).

3.2 Titration/Quantifying the Virus

1. Serial dilution of the virus needs to be performed prior to quantifying the virus using a plaque assay.
2. Vero cells need to be plated in a 6-well plate to 90% confluence prior to initiating this assay.
3. A $\log_{10}$ fold dilution of the virus should be performed (as shown in Fig. 1) in microcentrifuge tubes using OptiMEM media until a dilution of 10^{-8} is reached.
4. Starting at 10^{-3} dilution, 1 mL of the dilution should be transferred to the monolayer of cells after aspirating the growth media on the cells.
5. Incubate the infected plate of cells at 37 °C for 2 h.
6. Aspirate the existing media, wash with PBS twice and add 2 mL of methylcellulose-laden media per well to all the wells.
7. Incubate for 72 h or until plaques start to appear.
8. Plaques can be identified by the formation of small gaps in the cell monolayer.
9. Add 1 mL of methanol slowly to the corner of each well.
10. Incubate at room temperature for 15 min.

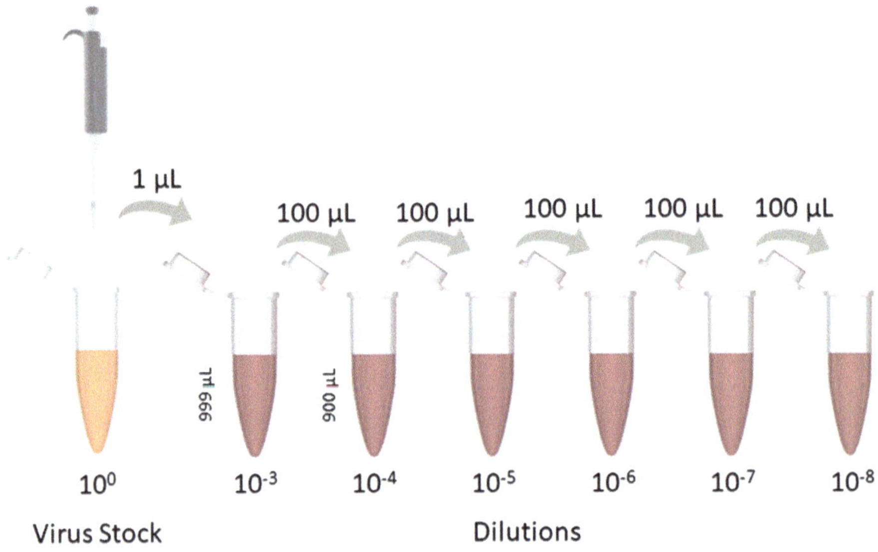

Fig. 1 Serial dilution for viral titration: From 1 mL of viral stock solution, transfer 1 μL of liquid to 999 μL of fresh OptiMEM solution to make a 10^{-3} viral dilution. Then, 100 μL of 10^{-3} dilution is transferred to 900 μL of fresh OptiMEM to make a 10^{-4} dilution. This process is repeated four more times to get a 10^{-8} dilution

11. Slowly aspirate the contents of the plate without disturbing the cell monolayer.
12. Add 1 mL of crystal violet working solution to each well.
13. Incubate the plate in dark for 30 min.
14. Discard the crystal violet solution and dry the wells on a sheet of absorbent paper.
15. Count the number of plaques at the highest dilution to quantify the total virus content in the starting solution.
16. Repeat the process twice prior to confirming the viral titer.

3.3 Preparing the Porcine Cornea for Infection [16]

1. After receiving the porcine eyes from the approved vendor, keep on ice if there is a delay in the processing of the tissue (as shown in Fig. 2).
2. Proper personal protection equipment needs to be used during this procedure to ensure protection from the aqueous humor or the other biological fluids spilling on to the investigator.
3. Spray a clean a working area with 70% ethanol to clean.
4. Spread a bench cover on the working area and tape the corners to ensure stability.
5. Take out the porcine corneas and put them in a 100-mm dish.
6. Gently wash the ocular surface with 1% AA solution in PBS.
7. Using 50-mm long gauze, firmly hold the posterior section of the eye in one hand.

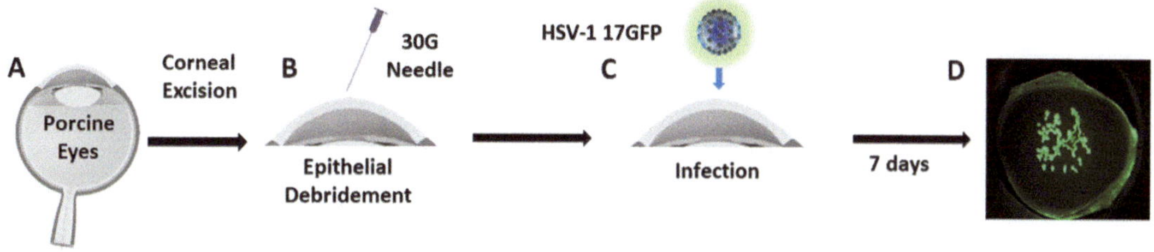

Fig. 2 Visualizing the corneal wounds generated by HSV-1 infection: (**a**) Freshly excised porcine corneal tissue. (**b**) Using a 30-G needle, make a single incision at the top of the epithelium. (**c**) Add 17 GFP virus in a volume of 1 μL to the corneal wound. (**d**) Incubate the porcine corneas for up to 10 days to visualize the dendritic spread of the virus over the cornea [17, 18]

8. This step can be done after the cornea is removed from the eye also; however, the process is much simpler when done while still attached to the eye.
9. Using a sharp blade, make an incision on the sclera at 1 mm distance from the cornea.
10. In a swift rotating action, cut around the edge of the cornea while ensuring that the vitreous humor does not spray out.
11. Using a flat-edged tweezer, hold one corner of the cornea-sclera edge and cut the remaining tethers to the eye using the blade.
12. Place the cornea in a 6-well plate overlaid with 2 mL of cornea media with epithelium facing up.
13. Using a 30-G needle, very gently make a single poke hole at the center of epithelial surface ensuring no damage to the stroma.
14. Add approximately 5 μL of virus solution accounting for 5×10^6 plaque-forming units (PFU) of gl86 KOS to the 3×3 epithelial debridement grid on the corneal surface (*see* **Note 9**).
15. Close the 6-well plate containing the corneas and place in incubator for 72 h.

3.4 Visualizing the Infection

1. Visualization of the virus should be done every day prior to the addition of the drugs.
2. Turn on the SteREO Discovery V20 microscope (Carl Zeiss, Germany, Jena, Germany) and LED light source.
3. Allow the lamp to warm up prior to imaging the corneas.
4. Without disturbing the cornea solution, carry the plate containing the corneas to the microscope.
5. Change the filter set to ensure that GFP (blue light) source is selected.

6. Modify the settings on the camera to capture the image with an exposure time of 500 ms.
7. Place the cornea plate under the stereoscope and capture the image of the entire cornea first at the lowest magnification (7.5×).
8. This image should be followed by a series of images with increasing magnification (e.g., 15×, 25×, and 35×) such that the viral spread and dendrite formation are clearly visualized.
9. Save all the images and carefully place the plate of infected corneas back in the tissue culture incubator.

3.5 Quantifying Infectious Virus Every Day

1. In order to understand whether the drug treatment on the cornea is working, virus titers from the porcine cornea should be evaluated every day (*see* **Note 10**).
2. One day prior to virus quantification, in a 6-well plate, seed Vero at a density of 5×10^5 cells per well.
3. Incubate overnight and ensure that a confluent layer of cells is present at the next day.
4. Pipette 500 μL of OptiMEM in to sterile microcentrifuge tubes.
5. In to each microcentrifuge tube, insert a sterile a Puritan Calgiswab with the cotton tip dipped into the OptiMEM solution.
6. Let the cotton swabs soak for a minimum of 5 min prior to use.
7. Carefully remove the plate containing infected porcine corneas from the tissue culture incubator without disturbing the underlying media.
8. Place the plate containing the corneas in the biosafety cabinet prior to opening the lid.
9. Use the pre-wetted Calgiswab to carefully swab the infected porcine corneal epithelium without causing any damage to the cornea.
10. Place the swab in the OptiMEM microcentrifuge tube and twirl it clockwise and anticlockwise five times.
11. Repeat **steps 9** and **10**.
12. Using a scissors, cut the metal tip of the Calgiswab short enough to allow the microcentrifuge lid to be closed.
13. Vortex the microcentrifuge tube for 1 min at high speed.
14. Perform the virus quantification via a similar protocol stated in Subheading 3.2.
15. In the virus dilution step, start the quantification at 10^{-1} and not 10^{-3} given that the total amount of virus is considerably lower on the porcine cornea.

16. Perform this quantification on days 1, 3, 5, and 7 post infection to assess the rate at which virus replicates in treated vs. nontreated porcine corneas.

3.6 Quantifying Viral Genomes Every Day [19, 20]

While the above-mentioned assay gives the quantification of available infectious titer, HSV-1 may also shed its virus in an asymptomatic fashion. In order to evaluate this, genome quantification can be performed from the samples collected from the porcine corneas.

1. Perform **steps 1–13** in the previous section.
2. Suspend 100 μL of the vortexed sample into 500 μL of virion buffer in a fresh microcentrifuge tube.
3. Add 1 μL proteinase K to this solution.
4. Incubate at 50–55 °C overnight.
5. Add 500 μL room temp ultra-pure phenol chloroform isoamyl alcohol mix (Ambion in culture fridge) to each sample and mix sample.
6. Centrifuge at 10,000 rpm (12,000 × *g*) 4 °C for 10 min.
7. Carefully move aqueous phase (top) to a fresh sterile tube.
8. Add 1 mL ice-cold 100% molecular grade ethanol to each sample's aqueous phase and mix.
9. Centrifuge at 10,000 rpm (12,000 × *g*) at 4 °C for 10 min.
10. Deoxyribonucleic acid (DNA) pellets may not be visible at this point; however, do not throw away samples.
11. Remove supernatant and wash pellet in 70–75% ethanol.
12. Centrifuge at 10,000 rpm (12,000 × *g*) 4 °C for 5 min.
13. Pellets should be more visible, but still may not be visible.
14. Remove supernatant and briefly dry pellet for no more than 20 min.
15. Resuspend in desired amount (10 μL) of molecular grade water.
16. Check quality and estimate DNA concentration by NanoDrop machine.
17. If multiple samples are present, equilibrate all the samples to have the same amount of total DNA by diluting with molecular biology grade water.
18. Prepare a stock solution per sample with SYBR green master mix as shown below.
19. After preparing the SYBR green master mix, aliquot the same into different microcentrifuge tubes.
20. Add requisite amount of DNA to each of the microcentrifuge tubes and label the tubes accordingly.

Table 1
Primer sequences for qPCR analysis: both porcine and HSV-1 primer sequences are given in this table

Porcine	
Glyceraldehyde 3-phosphate dehydrogenase (GAPDH)	F:TACACTGAGGACCAGGTTGTG R: TTGACGAAGTGGTCGTTGAG
Interferon (IFN)-β	F: ACTGGCTGGAATGAAACCGT R: AATGGTCATGTCTCCCCTGG
IFN-α	F: CCAACCTCAGCCTTCCTCAC R: TCCTCATTTGTGCCAGGAGC
Tumor necrosis factor (TNF)-α	F: GGCCCAAGGACTCAGATCAT R: CTGTCCCTCGGCTTTGACAT
HSV-1	
VP16	F: TCGGCGTGGAAGAAACGAGAGA R: CGAACGCACCCAAATCGACA
gD	F: TACAACCTGACCATCGCTTC R: GCCCCCAGAGACTTGTTGTA

21. Into a 384-well plate, add 8 μL of DNA SYBR green mix in triplicates for each primer. For example, if 2 primers are used, fill the sample in to 3 × 2 = 6 wells.
22. Thaw the working primer solutions for porcine GAPDH and HSV-1 gD at 4 °C.
23. For each sample, add 2 μL of primer in triplicate, that is, if 3 samples are present, add 2 μL of GAPDH primer and 2 μL of gD primer to each of the samples in triplicate 3 × 3 = 9 wells.
24. Cover the plate with a transparent PCR plate film tightly.
25. Centrifuge the PCR plate at 3000 rpm (1,000 × *g*) for 5 min.
26. Run a qPCR to detect the presence of viral genomes in the porcine corneal samples (Table 1).

3.7 Quantifying Host and Viral Transcripts

In order to understand whether the HSV-1 infection or the drug treatment caused significant differences to the porcine cornea, this protocol can be used by measuring the host interferon transcript levels.

1. Using a blunt blade, slowly scrape the corneal epithelium into 1 mL of fresh cornea medium.
2. Using a 100-μL pipette, vigorously mix the liquid to break any large chunk of cells present in the culture media.
3. Collect all the scraped cells in to a microcentrifuge tube.
4. Centrifuge the cells at 800 × *g* for 5 min.
5. Break the cell pellet by flicking the bottom of the tube gently with your finger.

6. Add 1 mL of TRIzol to the cell pellet and mix vigorously while breaking the pellet.
7. It is important to break the pellet to ensure that all the cells are being dissolved by TRIzol.
8. Add 0.2 mL of molecular biology grade chloroform to each of the samples.
9. Shake the samples vigorously for 20 s to ensure complete mixing of both the liquids.
10. Centrifuge the mixtures at top speed (14,000 × *g*) in the tabletop centrifuge for 10 min at 4 °C.
11. Collect the aqueous (top) layer in a fresh tube without disturbing the intermediary or bottom layers.
12. Add 0.5 mL of 100% 2-propanol to each of the aqueous layers.
13. Incubate at room temperature for 5 min.
14. Centrifuge at top speed for 10 min at 4 °C.
15. Discard the supernatant. The ribonucleic acid (RNA) pellet may or may not be visible.
16. Add 75% molecular biology grade ethanol to the pellet to precipitate the RNA.
17. Centrifuge at top speed for 10 min at 4 °C.
18. Discard the supernatant and allow the pellet to dry on ice for no more than 20 min.
19. Add 20 μL of molecular biology grade water to each of the RNA samples.
20. Quantify the RNA in all the samples using NanoDrop and equilibrate them with molecular biology grade water (*see* **Note 11**).
21. Using the manufacturer's protocol, reverse transcribe the RNA to cDNA (*see* **Note 12**).
22. Follow the protocol for qPCR mentioned in the previous section to assess transcript levels for each of the samples using the primer sequences mentioned in the reagents section (Table 2).

4 Notes

While most of the techniques mentioned in this chapter are straightforward, there are instances where the protocol mentioned might not yield the best of results. The following notes can help the investigator keep a look out for any possible problems associated with the above-mentioned techniques.

Table 2
Formula Table for qPCR: Please use this formula table to calculate the amount of SYBR Green mix, primer volume, and sample RNA to be able to perform a successful qPCR experiment

Variables	Formula	Sample
Number of samples	X	3
Number of replicates	3	3
Number of primers	Υ	4
Pipetting errors	$(X + 0.5)$ and $(\Upsilon + 0.5)$	3.5 and 4.5
SYBR Green Mix (μL)	$Z = 5 \times 3 \times (X + 0.5) \times (\Upsilon + 0.5)$	236.25
Mol. bio grade water	$Z/2$	118.125
DNA per sample	$0.5 \times (\Upsilon + 0.5) \times 3$	6.75
SYBR + water per sample	$(1.5 \times Z)/(X + 0.5)$	101.25

1. Do not use the working primer stocks for longer than a month or when you start observing multiple peaks in the melt curve. Make a fresh stock every time.
2. Always wear proper personal protection equipment such as a lab coat, gloves, eye-gear and mask while performing any of the protocols mentioned in this chapter.
3. Methyl cellulose needs to be autoclaved separately and cooled prior to adding the whole media to it. Otherwise, it might result in contamination or nonfunctionality of the media.
4. 5% Methylcellulose media takes a long time to dissolve, hence ensure that the media is prepared prior to starting the experiments.
5. Crystal Violet powder and stock solutions are carcinogenic and needs to be treated with care.
6. Crystal violet stains are hard to remove from clothes and hence much care needs to be taken.
7. Sonication needs to be performed inside a fume hood. This ensures that any virus that is aerosolized can be kept away from the investigator.
8. Autoclaved milk consists of 5% non-fat milk that is autoclaved 3 times in order to get a coffee brown color.
9. gL86 virus is slower to grow compared to other herpesvirus strains. Observe the cells every day to see if they circularize during infection.
10. The best viral titer can be achieved when the infected cells appear circular while still attached to the culture dish/flask.

11. Ribonucleic acid (RNA) and DNA analysis should always be performed while the samples are kept in cold temperatures. Having an ice bucket handy can help reduce the degradation of nucleic acids.
12. The High-capacity cDNA kit comes with buffers and dNTPs along with Reverse Transcriptase enzyme. The reverse transcriptase enzyme remains in liquid form even when stored at -20 °C freezer while the other reagents become solid. So do not take the enzyme out of the freezer while the other reagents thaw to room temperature. This ensures longer life for the enzyme.

Acknowledgements

This work was supported by grants from the NIH (R01EY024710-05, R01EY029426) to D.S.

References

1. Liesegang TJ (2001) Herpes simplex virus epidemiology and ocular importance. Cornea 20 (1):1–13
2. Farooq AV, Shukla D (2012) Herpes simplex epithelial and stromal keratitis: an epidemiologic update. Surv Ophthalmol 57 (5):448–462. https://doi.org/10.1016/j.survophthal.2012.01.005
3. Farooq AV, Valyi-Nagy T, Shukla D (2010) Mediators and mechanisms of herpes simplex virus entry into ocular cells. Curr Eye Res 35 (6):445–450. https://doi.org/10.3109/02713681003734841
4. Farooq AV, Shah A, Shukla D (2010) The role of herpesviruses in ocular infections. Virus Adapt Treat 2(1):115–123
5. Xu F, Schillinger JA, Sternberg MR et al (2002) Seroprevalence and Coinfection with Herpes Simplex Virus Type 1 and Type 2 in the United States, 1988–1994. J Infect Dis 185(8):1019–1024. https://doi.org/10.1086/340041
6. Xu F, Sternberg MR, Kottiri BJ et al (2006) Trends in herpes simplex virus type 1 and type 2 seroprevalence in the United States. J Am Med Assoc 296(8):964–973. https://doi.org/10.1001/jama.296.8.964
7. Koganti R, Yadavalli T, Shukla D (2019) Current and emerging therapies for ocular herpes simplex virus type-1 infections. Microorganisms 7(10):E429
8. Lobo A, Agelidis AM, Shukla D (2019) Pathogenesis of herpes simplex keratitis: the host cell response and ocular surface sequelae to infection and inflammation. Ocular Surface 17 (1):40–49
9. Koujah L, Suryawanshi RK, Shukla D (2019) Pathological processes activated by herpes simplex virus-1 (HSV-1) infection in the cornea. Cell Mol Life Sci 76(3):405–419
10. Lass JH, Langston RH, Foster CS et al (1984) Antiviral medications and corneal wound healing. Antivir Res 4(3):143–157
11. Burns WH, Saral R, Santos GW et al (1982) Isolation and characterisation of resistant Herpes simplex virus after acyclovir therapy. Lancet 1(8269):421–423. https://doi.org/10.1016/s0140-6736(82)91620-8
12. Crumpacker CS, Schnipper LE, Marlowe SI et al (2010) Resistance to antiviral drugs of herpes simplex virus isolated from a patient treated with acyclovir. N Engl J Med 306 (6):343–346
13. Yildiz C, Ozsurekci Y, Gucer S et al (2013) Acute kidney injury due to acyclovir. CEN Case Rep 2(1):38–40. https://doi.org/10.1007/s13730-012-0035-0
14. Fleischer R, Johnson M (2010) Acyclovir nephrotoxicity: a case report highlighting the importance of prevention, detection, and treatment of acyclovir-induced nephropathy. Case Rep Med 2010:1–3. https://doi.org/10.1155/2010/602783
15. Thakkar N, Jaishankar D, Agelidis A et al (2017) Cultured corneas show dendritic spread and restrict herpes simplex virus infection that

is not observed with cultured corneal cells. Sci Rep 7:42559
16. Heichel J, Wilhelm F, Kunert KS et al (2016) Topographic findings of the porcine cornea. Med Hypothesis Discov Innov Ophthalmol 5 (4):125–131
17. Duggal N, Jaishankar D, Yadavalli T et al (2017) Zinc oxide tetrapods inhibit herpes simplex virus infection of cultured corneas. Mol Vis 23:26–38
18. Yadavalli T, Agelidis A, Jaishankar D et al (2017) Targeting herpes simplex virus-1 gD by a DNA aptamer can be an effective new strategy to curb viral infection. Mol Ther Nucleic Acids 9:365–378
19. Berger A, Preiser W (2002) Viral genome quantification as a tool for improving patient management: the example of HIV, HBV, HCV and CMV. J Antimicrob Chemother 49 (5):713–721. https://doi.org/10.1093/jac/dkf050
20. Mackay IM, Arden KE, Nitsche A (2002) Real-time PCR in virology. Nucleic Acids Res 30 (6):1292–1305

INDEX

A

B

Hiranmoy Das (ed.), *Wound Regeneration: Methods and Protocols*, Methods in Molecular Biology, vol. 2193, https://doi.org/10.1007/978-1-0716-0845-6,

D

E

F

G

Q

R

MIX
Papier aus verantwortungsvollen Quellen
Paper from responsible sources
FSC® C105338

If you have any concerns about our products,
you can contact us on
ProductSafety@springernature.com

In case Publisher is established outside the EU,
the EU authorized representative is:
Springer Nature Customer Service Center GmbH
Europaplatz 3, 69115 Heidelberg, Germany

Printed by Libri Plureos GmbH
in Hamburg, Germany